PEOPLE'S SCIENCE

PEOPLE'S SCIENCE

BODIES AND RIGHTS ON THE STEM CELL FRONTIER

RUHA BENJAMIN

STANFORD UNIVERSITY PRESS • STANFORD, CALIFORNIA

Stanford University Press
Stanford, California

Janus illustration by John Alan Birch. Reproduced by permission.

Sankofa Bird illustration by Kenneth Anthony Burney. Reproduced by permission.

Printed in the United States of America on acid-free, archival-quality paper.

Library of Congress Cataloging-in-Publication Data

Benjamin, Ruha, author.
  People's science : bodies and rights on the stem cell frontier / Ruha Benjamin.
      pages cm
  Includes bibliographical references and index.
    ISBN 978-0-8047-8296-8 (cloth : alk. paper) -- ISBN 978-0-8047-8297-5 (pbk. : alk. paper)
  -- ISBN 978-0-8047-8673-7 (electronic)
      1. Stem cells--Research--Social aspects--California. 2. Stem cells--Research--Government
  policy--California. 3. Embryonic stem cells--Research--Social aspects--California.
  4. Embryonic stem cells--Research--Government policy--California. 5. Medical policy--
  Social aspects--California. I. Title.
      QH588.S83B46 2013
      616.02'774072--dc23

                                2013006484

Typeset by Bruce Lundquist in 10.5/15 Adobe Garamond

for behin & truitt

We seem to be made to suffer. It's our lot in life.

—*C-3PO*, Star Wars *(1977)*

There has to be more that we can do, a better destiny that we can shape.
Another place. Another way. Something!

—*Octavia Butler,* Parable of the Sower *(1993)*

In the technical realm, we repeatedly enter into a series of social contracts, the
terms of which are revealed only after the signing.

—*Langdon Winner,* The Whale and the Reactor *(1989)*

What we need today, for the sake of the survival of this planet, is long-term
vision. Can governments whose very survival depends on immediate, extractive,
short-term gain provide this? Could it be that democracy, the sacred answer
to our short-term hopes and prayers, the protector of our individual freedoms
and nurturer of our avaricious dreams, will turn out to be the endgame for the
human race?

—*Arundhati Roy, Harvard STS Science and Democracy lecture (2010)*

CONTENTS

PREFACE

As a social scientist interested in learning how science and technology can serve as a window onto broader questions of health, equity, and justice, I have often felt pressured to bracket my personal experiences so as not to "taint" my analytic gaze. This, despite the fact that those we might call *non*–social scientists, who practice the "hard" disciplines, are socialized within particular cultural milieus and known to bring their personal biographies into the lab.[1] I inevitably still struggle to study the social world while caring deeply about what kind of world we bring into being, and so I have worked to sharpen an objectivity grounded in self-reflection, one by which I take stock of my experiences, assumptions, and commitments so that I can produce a more complete and contextualized representation of my research subject.

As life would have it, a week before the final revisions of this manuscript were due, my father was rushed to the hospital with strokelike symptoms. Having already undertaken the massive task of learning to walk and work again after his left side was paralyzed due to a stroke three years ago, *here he was again*: in a small hospital room at Cedars-Sinai, where neurologists were trying to figure out an effective course of treatment. This time around there was one crucial difference. After his first stroke, my dad's employer-based health insurance covered the multiple tests, treatment, and physical therapy that allowed for his slow recovery. But this time, working in a new and precarious line of work, he was uninsured. So the longer, more substantial, and perhaps more effective his treatment turns out to be, the bigger and more burdensome the bill that awaits him when he eventually checks out of the hospital.

There is little doubt that this and similar firsthand experiences, where I have been confronted by the coexistence of abundance and scarcity, form the invisible dark matter that holds these pages together. Social principles in this way are the corollary of scientific theories—by putting them aside we are choosing not to see important features of how reality takes shape. Like those who investigate dark matter itself, our evolving understanding is not an inevitable outcome of simply using particular tools.[2] Instead, our *ideas* about the relationship between different parts of the universe, our refined assumptions, make it possible to detect forces that were completely invisible before. The more we are attentive to our theories and principles, the better our research; not only is our capacity for producing data enhanced but hopefully so is our ability to revise our assumptions when confronted with the unexpected. A social principle in this way can and should share the flexibility of a theorem.

This brings us to one of the core themes of this book: in a society of haves and have-nots, the biomedical ingenuity that can bring people back from near death has a *double edge*—one side excising brain embolisms from blood vessels where they do not belong, and the other side deepening the fault lines through which our current social order distributes suffering and premature death in radically uneven ways. As Jo Phelan and Bruce Link put it, "[w]hen we develop the ability to control disease and death, the benefits of this new-found ability are distributed according to resources of knowledge, money, power, prestige, and beneficial social connections."[3] But does this mean biomedical ingenuity must *necessarily* deepen inequalities? Before undertaking this study, I assumed so, in part because I experienced science and medicine primarily as an outsider and was not aware of the openings through which innovation and equity could be mutually constructed.

My experiences as an "outsider within" tell me we don't have to put scientific innovation on hold for more equitable social relations to first take shape.[4] Instead, we are in dire need of bolder visions of innovation that are seeded with a commitment to invigorate more equitable social relations *alongside* the pursuit of regenerative medicine and other fields. This necessitates that conditions we already know to foster health and well-being are not taken for granted in the process. Drawing attention to the "shadow realities"[5] of those who do not have access to affordable healthcare will, for example, make it apparent that a more varied people must participate in the governance of science, technology, and medicine, fully contributing to the decisions that impact their lives and, even more fundamentally, their life chances.

As an African American woman of mixed ethnic heritage and someone who specializes in qualitative research methods that require a fair amount of interpretation, my precarious location in the academy has heightened my attention to this question of how experiences shape knowledge production. I have walked a fine line in the pages to come, seeking to represent the social and political struggles over new biotechnologies with *commitment* to the principles of equity and justice I hold dear, and with *fairness* to my research subjects. Some readers may come to question my evenhandedness, wishing I had infused less assessment and interpretation of the stakes of these struggles into this work, while others will wonder why I did not come down harder on one side or another. To both I acknowledge, the line I walk is more like San Francisco's Lombard Street. Crisscrossing the landscape of scientific empiricism and cultural meaning, my aim is less to strike a perfect balance *or* arrive at a final destination than to introduce those I have met along the way to others whose social worlds they may have ignored, dismissed, or misunderstood.

ACRONYMS

ART: assisted reproductive technology

CGS: Center for Genetics and Society

CIRM: California Institute for Regenerative Medicine

ELSI: ethical, legal, and social implications

FUNDAEC: Fundación para la Aplicación y Enseñanza de las Ciencias (Foundation for the Application and Teaching of the Sciences)

ICOC: Independent Citizens' Oversight Committee

iPSCs: induced pluripotent stem cells

IVF: in vitro fertilization

NAAPC: National Association for the Advancement of Preborn Children

NYSTEM: New York Stem Cell Science

SCI: spinal cord injury

SCNT: somatic cell nuclear transfer

SCR: stem cell research

PEOPLE'S SCIENCE

INTRODUCTION

To the Moon

Before we start designing ways to get to the moon, can we just make
sure everybody on my block can actually get to work?
—*Patricia Berne, scholar-activist*[1]

The fear of imagination in politics comes from the fear of illusion. It is like
refusing to use a tool at all because it can be misused.
—*Richard Sennett,* Authority[2]

ON FEBRUARY 16, 2007, I sat in a San Francisco Sheraton meeting room waiting for then California governor Arnold Schwarzenegger to appear. The
governing board of California's new stem cell agency had scheduled a
press conference for the governor to announce his approval of a $150 mil-
lion loan to fund the first round of scientific grants to stem cell researchers. This move by the governor was politically significant, according to
Donna Gerardi Riordan, director of programs at the California Council
on Science and Technology, "because it occurred one day after President
Bush vetoed bipartisan legislation that would have relaxed federal restrictions on stem cell research."[3] The loan was necessary after $3 billion in
state bonds that should have been available to fund the agency's work
were held up by two lawsuits arguing on pro-life or consumer rights
grounds against the California Stem Cell Research and Cures Initiative.

Having attended a number of the stem cell agency's public meetings,
I was routinely mistaken for a journalist as I hurriedly typed my field
notes; so on this occasion I was ushered into the front row of the press
conference, several feet away from the podium. By the time the governor arrived, leaning on a cane on account of a skiing accident, the room

was abuzz with anticipation. He congenially shook hands with about a dozen of the board members and patient advocates standing at attention behind the podium, his limp in sync, as it were, with the impairments of those advocates placed front and center in wheelchairs, wearing their "Stem Cell Action Network" buttons. As the buzz turned to a hum and the hum gave way to absolute quiet, the governor began:

> These initial grants today are very important because you all know we cannot afford to wait when it comes to advancing life-saving science. So today is a day of great hope. We have hope for promise of incredible advances in medicine. Hope for the eventual end of suffering from diseases like Alzheimer's, Parkinson's, cancer, and MS and hope for the people who love someone with one of those terrible diseases. . . . I know a lot of people in California and around the world that have diseases like that and can be helped with this important research. This is why we are not waiting for anyone to do it for us. We are creating the action right here in California [audience applause].
>
> I also want to show my deepest gratitude to the scientists and to the doctors who are receiving this money to find new therapies and new cures. I just want them to know that I am 100 percent behind you and the people of California are 100 percent behind you. We think the world of them. They are opening up possibilities that only a few years ago, we would have only imagined. So they are our newest action heroes [audience laughter], and I am looking forward to what they can achieve.[4]

Strategically distancing himself from an unpopular U.S. president who had just restricted the work of stem cell researchers, the governor effectively invoked the hope and heroics that have animated this new field since the isolation of human embryonic stem cells in 1998. But whereas the governor's comments focused on the heroics of scientists, it was the actions of the advocates, policy entrepreneurs, lawyers, and journalists in the room that in fact had enabled passage of the stem cell initiative—and the scientists were beholden in many ways to these various constituencies and their sometimes competing agendas. This shifting relationship between science and society is what *People's Science* sets out to explore, revealing those struggles, both manifest and veiled, that animate science shaped by public demands.

The limits of the old trickle-down relationship between science and society, in which the public was expected to patiently wait for the fruits of science to reach it,[5] first grew apparent in the wake of the 1945 atomic bomb tests. Since then, the controversy over genetically modified organisms that started in the 1970s and continues today; the revelations of the Tuskegee syphilis trials; and reports of numerous medical abuses against women, racial and ethnic minorities, people with disabilities, indigenous populations,[6] and prisoners,[7] together reveal the underside of scientific innovation. Greater awareness that these are not fringe events carried out by hacks and mad scientists but often take place in mainstream institutions, frequently at the direction of the most prominent members of their fields, has fueled a growing movement against allowing researchers to govern themselves, in isolation from wider social norms.

But what some consider the "contamination" of pure science is not only a bottom-up process of citizens demanding more input and regulation.[8] The commercialization of scientific research has also increased exponentially. The federal Bayh-Dole Act was passed in 1980 to encourage the more efficient commercialization of research, thereby incentivizing the pursuit of profit on the part of researchers. The rapid growth of patent licenses and the widespread development of technology transfer offices at public institutions, having a mandate to identify research ideas that could be developed into commercial products, raise new concerns about the conflicts of interest that frequently inform scientific investigation. In a landmark case that some call the "Brown v. Board of Education for genetic science,"[9] the U.S. Court of Appeals for the Federal Circuit "affirmed the right of Myriad Genetics to patent two genes linked to breast cancer [BRCA-1 and -2] overturning a lower court ruling that threatened a key element of the biotech business."[10] Critics of gene patenting say it creates a monopoly that impedes research "that could lead to better diagnostic treatments" and prevents competition, which ultimately forces consumers to pay more and limits women's treatment options. "Tests that cost $300 end up costing $3,000 because of the patent monopoly that the company has," said Lori Andrews, a law professor at Chicago-Kent College of Law.[11] By contrast many in the biotech industry are relieved by the ruling: Goldman Sachs analyst Isaac Ro said the Myriad ruling lifts a "near-term threat to investors," even as he and others expect the legal fight to continue.[12] The life sciences and the burgeoning biotech industry are especially vulnerable to conflicts between commercial, medical, and broader social interests, as the application of commercial logic to (and commodification of) the human body leads us full circle to the dangerous medical practices of World War II—and even prior to that, to American chattel slavery. Together, these ethical and commercial concerns have laid to rest the figure of the disinterested and autonomous scientist traversing what Vannevar Bush, in a Report to the President, once called the "endless frontier" of knowledge.[13]

But while some observers worry that the cloistered and disinterested ideal has been replaced by self-interested, profit-seeking scientist-entrepreneurs,[14] evidence suggests that many scientists are taking a longer view. A long list of eminent scientists spoke up in support of the American Civil Liberties Union case against the Myriad patents. In less litigious contexts, some are going so far as to leverage the far-reaching implications of their work to resonate with a broad public mandate of health access and democratic inclusion.[15] They are supplanting the trickle-down ideal with a participatory ethos, in name if not in practice, in which they engage with nonscientist stakeholders at ever earlier stages of the research and development process. Often this engagement is geared toward supportive patient advocates who provide the moral imperative for rapid scientific development; but sometimes other constituencies manage to get the attention of the scientific establishment.

Indeed, the giving or withholding of public approval has grown more direct: in over a dozen U.S. states, initiatives to fund or ban stem cell research have come to a popular vote. California's Stem Cell Research and Cures Initiative was one such effort, in which scientists, policy entrepreneurs, and patient advocates worked together to achieve unprecedented investment in and insulation for the besieged new field. On November 2, 2004, Proposition 71 successfully passed, following a massive "pro-cures" campaign that linked investment in the new science with the alleviation of suffering from over eighty diseases. It authorized the sale of state bonds in the amount of three billion dollars over ten years, to be managed by a new stem cell agency (the California Institute for Regenerative Medicine, or CIRM), which would be protected by an amendment to the state constitution that created a "right to research." The text of the initiative also stipulated that the new agency was to be governed by an Independent Citizens' Oversight Committee (ICOC), the composition of which has proven an ongoing locus of struggle over the legitimate parameters of public inclusion and representation.[16]

Compared with the old, cloistered model of science, the initiative and its governing structure appear radically inclusive: approval is by the voters, funding by taxpayers, and governance by representatives of the public. Yet it still does not fully address the concerns and expectations characteristic of the participatory trend in science development. Many of the board's critics point to the economic and institutional conflicts of interest that may cloud members' ability to implement Prop. 71. (One such conflict forced the resignation of a board member who held stock options in a company that had applied for a grant from the stem cell agency.) Beyond this, however, I suggest that a *lack* of constructive conflict over the priorities and governance of science poses an even more fundamental challenge to a truly participatory initiative such as this. The lack of public accountability to, and inclusion of socially subordinate collectives is, in my estimation, more politically worrisome and sociologically interesting than is the stain of stock options. Without this deeper accountability, proponents of a "right to research" are sacrificing social equity at the altar of scientific expedience.

The lack of robust deliberation about how scientific initiatives can and should reflect a wider array of social concerns is due in part to the systemic exclusion of those who could articulate concerns about state investment in stem cell research from working-class, feminist, disability, or racial justice points of view. But these omissions also stem from a fundamental ambiguity about who "the people" of participatory science initiatives are and should be. In focusing on how "the people" of a people's science are constructed and contested, I offer a critical understanding of the processes of inclusion and exclusion that typically remain hidden in depictions of "the stem cell debate."

These struggles over the credible parameters of involving people's bodies *and* interests in stem cell research are fundamentally different from nonscientific political struggles, because the question of what the state owes particular groups is intimately connected to biological definitions of *what constitutes a group* in the first place. The coemergence of novel life sciences and new rights claims that "redefine the obligations of the state in relation to lives in its care" is what Harvard professor Sheila Jasanoff terms *bioconstitutionalism.*[17, 18] We see glimpses of bioconstitutionalism in the now codified "right to research" brought about by the passage of Prop. 71. But we also find it in the pro-cures assertion that it is a right of families to pursue the best course of treatment for their loved ones despite the ethical toes such treatment steps on. This relationship between a "right to research" in a laboratory and sociopolitical rights in the political arena does not pertain simply to the realm of official policy and legislative enactment. Rather, we find it in *bioconstitutional moments*, where struggles over who *we* are, what we are owed, and what we are responsible for, as both objects and subjects of scientific initiatives, are taking place all around us. In California and a growing number of jurisdictions, representatives of various constituencies are attempting to codify answers to these questions.[19] In an even greater number of arenas, people have yet to formalize answers but are tinkering with and sometimes brawling over the role and interests of the public in conducting controversial science.[20]

A political sociology of science requires that we examine not only courts, legislative sessions, board meetings, and ballot boxes but also funding agencies, hospital clinics, and other, more mundane sites where the meaning of life and the entitlements owed to the living are negotiated and contested. In this way, the seeming exceptionalism of California gives way to an expanded social terrain with a common propensity to struggle over boundaries of inclusion. It is upon such fractured ground, and not upon any firm authority and hegemony on the part of science *or* overwhelming trust or consent on the part of society, that public engagement with science is taking place. People do not simply "hold" stakes but actively construct and calibrate the risks of aligning their interests with scientific initiatives, forming the supple social infrastructure of stem cell research and related life sciences—what one racial justice advocate working on organizing a coalition to demand greater inclusion for minority health interests in the California stem cell initiative called a "house of cards."[21]

In contrast to the polarizing frames of Right-Left politics, so much of the actual work to advance or oppose scientific research is carried out via politically promiscuous bedfellows. We find a pro-choice alliance teaming up with the Catholic Church to object to the use of oocytes for research; a sickle cell disease organization signing on to a stem cell campaign only to be formally excluded from the initiative's implementation; and conservative activists such as Mel Gibson speaking out against state investment in "unethical experimentation" to a predominantly working-class African American community in Watts, Los Angeles, whose local health clinic had closed on account of budget cuts. Such novel alliances and collaborations are indicative of the way in which a controversial scientific field does not simply fall along old sociopolitical boundaries but redraws them in unpredictable ways.[22]

Consider one such foray into this shaky social terrain. In the fall of 2006, Dr. Zach Hall, former National Institutes of Health director and president of the new California stem cell agency, found himself in front of an unlikely audience. Nearly two years after the historic passage of Proposition 71, Hall was invited by members of the Oakland-based Black Wall Street Merchants Association and the Black Board of Trade and Commerce to participate in a "two-way dialogue" about the significance of stem cell research for African Americans. The ensuing town hall-like forum, with over two hundred physicians, lawyers, clergy, heads of social service agencies, and community college and public school administrators in attendance, was eventually televised on a local station. Hall fielded questions about whether diseases affecting the African American community would be prioritized by stem cell researchers; how small black-owned firms could compete against more established companies for grants; and strategies for African American students and scientists to be drawn into the stem cell career pipeline.

Even as forum attendees expressed a strong interest in the wider social and economic impacts of stem cell research, Hall attempted to gently resist these larger public considerations. Presentation slides and luncheon plates in place, Hall explained that the official text of the stem cell proposition did not in fact entail the social inclusion priorities that attendees raised. He also hinted that it was not realistic for the small staff of fifty people at the new stem cell agency to prioritize early-stage research based primarily on its impact on a particular population or tackle such deeply entrenched problems as the unequal availability of science education. Acknowledging the importance of such endeavors without accepting responsibility for advancing them, he carefully tried to hold together the fragile bond symbolized by Prop. 71. He did so by staying as close to the science as possible and then astutely drawing participants' attention to the problematic politicization of science by the opponents of stem cell research. After explaining the basics of the field, Hall emphasized the importance of a diverse pool of tissue donors to ensure the future applicability of stem cell treatments to African Americans, noting that

> [i]f stem cells are to be useful for all members of our population, we want stem cells that reflect in their genetic characteristics all the diversity found in the human population, and the problem with in vitro fertilization clinics, as somebody said, it's a very limited population. They have to be rich, white, and infertile. And we need more stem cell lines than that. We need stem cell lines of all sorts, of all sorts of people.[23]

In effect, Hall resisted the role of populist scientist being thrust upon him even as he carved out a much more biologically circumscribed understanding of how scientists could ensure a science "for the people." He proffered a kind of inclusion based on the presumed genetic diversity of different races, offered as evidence of the agency's forward-thinking agenda, even as he relied on an older notion of biologically based racial differences as the primary basis on which the initiative was prepared to consider the inclusion of African Americans.

Remaining noncommittal with respect to the various visions of social inclusion offered by the attendees, ideas that he promised "to take back to the board," Hall instead carefully weighed in on the major fault line then dominating the stem cell terrain, namely the moral status of embryos. He explained that one of the groups suing the new agency called themselves the National Association for the Advancement of Preborn Children (NAAPC), inverting the last two letters of the well-known civil rights organization, the National Association for the Advancement of Colored People (NAACP). Audience members let out a knowing groan at this appropriation, whereupon Hall noted that the NAAPC's lawsuit was on behalf of an unborn child named "Jane Scott Doe"—the plaintiff's middle name evoking that of Dred Scott, the African American slave who unsuccessfully sued for his freedom in 1857.[24]

Betting that his audience would find the conflation of American chattel slavery with the use of embryos in research disingenuous if not offensive, Hall successfully drew attention away from his noncommittal stance on the social inclusion issues raised at the Oakland forum, and toward the machinations of the NAAPC. In focusing on the problematic politicization of science on the part of the agency's adversaries, Hall implicitly cautioned his audience that such political contamination would slow the development of cures. In so doing, he never had to say that the participants' demands for a broader social commitment posed a similar threat.

When I sat down with the president of the Black Wall Street Merchants Association, Eddie Dillard, to discuss the accountability of the fledgling stem cell agency, he was adamant that "if we're gonna pay, we need to play!" even as he hedged about what "playing" in this speculative terrain would actually look like. He and others on the front lines of advocating inclusion for racial and ethnic minority, feminist, and disability concerns in publicly sponsored science initiatives seek to challenge the invisible interests of the amorphous "public" to which Proposition 71 was pitched, though they do not necessarily have a full-blown alternative to propose. Even so, they appear to understand that accountability cannot be achieved in an afternoon dialogue or a onetime Q&A session but only via an ongoing and fully institutionalized means of ensuring that in the case of stem cell research, social goals, and not only biomedical goals, are addressed. In the words of one of the more outspoken racial justice health activists, Joseph Tayag, ethnoracial minorities "need to be *at* the table, not just *on* the table" of stem cell research.[25]

This Oakland forum, in which participants sought to actively construct, not simply "hold," specific stakes in stem cell research, exemplifies the new contract between science and society that is being negotiated in California and elsewhere. Confronted with the complexity of implementing an initiative like Proposition 71, its architects are faced with the fragility of their pact with the public. Amid the great diversity of the electorate, whose interests should take precedence? Through what channels should they be heard? By what standards should science be held responsible? In short, now that the science "for the people" rhetoric had achieved its purpose, winning onetime approval at the ballot box, how would consent be maintained? How would accountability be enacted and, in time, contested?

These and more questions not only animated the Oakland stem cell forum but are part of a larger process of political experimentation in which the parameters of social inclusion in science are being redrawn. The issues raised and dodged at the Oakland luncheon, in terms of which scientists and various publics typically negotiate the broader stakes of scientific investment, have figured in public debate in over a dozen states that have used electoral and legislative processes to fund or ban stem cell science. At the same time, the United States lags far behind the participatory mechanisms well underway outside of the country: "science shops," in which researchers collaborate with citizens; "science courts," in which laypeople pass judgment on scientific controversies; "citizen boards" to assess technological risks.[26] In all these efforts, scientific norms still tend to outweigh wider social norms in terms of quality control and with respect to measuring success. Examining what gives these hybrid political-scientific experiments legitimacy and staying power—whether the authority of science, populist exuberance, or as I argue, the strategic fabrication and mobilization of a particular kind of consenting public—is one of the tasks set forth in this book.

To the Moon?

In 1998, University of Wisconsin developmental biologist James Thomson announced that his lab had managed to isolate and culture human embryonic stem cells from the inner lining of a human embryo, a feat achieved until then only in animals.[27] Soon after, frenzy ensued. As one commentator predicted, "the stage was set for a raging battle in which scientists, politicians, religious leaders, doctors, and patients would find themselves unwilling soldiers."[28] For many opponents of the technique, the potential benefits of such research were offset by the ethics of sacrificing what they regarded as a potential person in the process. For many who believe that human life begins at conception, the cost of using this technique was too high. Supporters of stem cell research, by contrast, weighed the possibility of relieving human suffering against such considerations. For them, the status of an eight-day-old embryo (blastocyst) was qualitatively less certain than that of living, breathing human beings who could benefit from regenerative medicine.

The issue of potentiality—potential humans and potential cures—pulls at both ends of the stem cell debate, impacting those not only in the research laboratory but in the political arena as well. The divergent hopes and fears surrounding the tools we use for both scientific and political experiments remind us how "the politics of biotechnology serves as a theater for observing democratic politics in motion."[29] Owing in large part to the technical advances of Thomson and colleagues, stem cell research has grown to be a wedge issue that both the political Right and Left use to cast the other side as enemies of life. Depending on how exactly one frames Thomson's achievement, the new field reflects mankind's ingenuity—or its fall from grace.

So when on August 10, 2001, President George W. Bush announced that National Institutes of Health grants could be used for research only on existing stem cell lines, numbering in the dozens, and that research producing new stem cell lines after that date would be ineligible for federal support—and would be illegal if done with federal dollars—sparks flew. Stem cell researchers and their supporters turned to the state initiative process, in which the electorate votes directly on a proposed law, as a political tool to push back restrictions against these controversial scientific techniques. The actual goal of California's Proposition 71 was to insulate the science of stem cell research from impinging political values, not to enhance public participation in the process—this despite the science "for the people" framing of the issue.

The pro–stem cell movement that was energized soon after Bush announced his restrictions climaxed on November 2, 2004, with the passage of Prop. 71 in California. But despite an extremely well funded campaign that flew in the face of federal limitations, not everyone was buying what they dubbed "21st century snake oil."[30] As longtime disability rights and racial justice activist Patricia Berne asked rhetorically, "Before we start designing ways to get to the moon, can we just make sure everybody on my block can actually get to work?" This and similar queries challenged both the funding and research priorities of the initiative and questioned in whose interests both were being advanced. In the estimation of Berne and others, stem cell research is akin to jet-setting to the moon while people's basic needs are going unmet.

Indeed, the massive state investment came at a time when a swelling state budget deficit had reached $16 billion[31]; the state had one of the highest housing foreclosure rates in the nation; over 20 percent of the population had no health insurance[32]; and the income gap was growing rapidly.[33] The socioeconomic disparities that such figures illustrate, however, were strategically overshadowed by advocates who linked stem cell investment to the pioneering ethos of the state:

> We created the fifth largest economy in the world, with a gross state product of more than $1 trillion, while holding sacred our commit-

ment to the stewardship of the land and its natural resources. We are quick to embrace new ideas and new citizens. Californians are dynamic, adaptable, and always focused on the future. Our state is the birthplace of the Mars Rover, the Digital Age and the Biotechnology Revolution.[34]

When, in 2001, President Bush restricted the use of stem cell lines in research that many posited as the future of medicine, it thus followed that California would offer itself as the place to resist his backward pull. In this vigorous reassertion of its futurist predilections, and against the backdrop of state economic crises, Californians made a high-stakes gamble: Prop. 71 passed while at the same time a ballot measure to extend basic healthcare coverage to the majority of its working classes who were then un- or underinsured (Proposition 72) was rejected. Indeed, proponents of the new field drew upon a ready futurist idiom that, by definition, obscures features of the present that systematically prevent health and wealth gains from "trickling down" to the masses.

Instead, California took a place among national governments as colleague and competitor. In fact, as one policy report indicates,

> even the federal support for stem cell research [was] dwarfed by the $295 million a year guaranteed by Proposition 71. In 2003, the National Institutes of Health awarded $24.8 million to researchers over the entire nation for embryonic stem cell research. . . . Proposition 71 created a source of grants that is nothing less than another NIH. . . . The United Kingdom, Singapore, Israel, and South Korea are often referred to as countries most supportive of stem cell research, but none have committed resources comparable to those promised in Proposition 71.[35]

Not surprisingly, then, the International Stem Cell Forum, a multilateral collaboration of seventeen entities that was set up in 2003, soon invited the nascent California Institute for Regenerative Medicine to join its ranks. In response, CIRM president Zach Hall exulted that the "inclusion of the state of California, along with national research organizations, demonstrates international recognition of the important role of CIRM in stem cell research."[36] Inducted alongside the Chinese Academy of Sciences and the University of Milan, California's stem cell agency established the state as an autonomous regulatory arena that was not wholly bound by the restrictive science policies of the federal government.

Lest this appear to be a story of Californian exceptionalism—cowboy science riding into the sunset of medicine's frontier—the electorate's investment in stem cell research is better seen as a byproduct of a broader pattern of the privatization of social goods, such as healthcare and education. Even before the passage of Prop. 71, almost all states offered tax breaks for the biotechnology industry, notably, to offset the financial losses endemic to this sector.[37] Nurturing the market and boosting scientific enterprise become two sides of the same shiny coin, as more and more of that enterprise relies on the private sector to translate basic knowledge into clinical goods. But what are the practical consequences in the context of stem cell research investment of boosting corporate welfare while the social safety net for the majority of people grows weaker? Stanford law professor Henry Greely and colleagues explain that

> Any new [stem cell] therapies would presumably be commercialized, and at that point market forces will play a large role in determining winners and losers. Sellers of new therapies will tend to benefit, and those who pay for them and those who sell current therapies that might be replaced would tend to lose. . . . There has been increasing awareness of the fact that many existing beneficial healthcare interventions are often not made equally available to all members of society. If new stem cell–based therapies require significant out-of-pocket spending, those with more discretionary dollars will be better positioned to benefit. . . . If in passing Proposition 71, voters envisioned distribution of new therapies according to medical need, rather than economic status, such unequal redistribution would be incongruent with their motivation.[38]

But in prioritizing scientific expedience over social equity, proponents of Prop. 71 discount the relationship between their exuberance and the creation of Greely's stem cell "losers."

While California elites were disproportionately represented at the polls,[39] it is important to realize that the lopsided investment in speculative versus basic healthcare was not fully class-specific, in large part owing to the effective packaging of Californian exceptionalism by the "Yes on 71" campaign. Rather than being invited to judge the proposed $3 billion investment within the context of socioeconomic disparities and the state's budget crisis, voters were offered a "prepackaged problematic" that cast the Bush administration's stem cell restrictions as constituting the *real* threat to Californians.[40] The autonomy sought by the state and the private sector was mirrored in the rhetoric of stem cell advocates, whose pursuit of biomedical cures drew upon the quintessentially American ideal of private self-determination. As Prop. 71 architect Robert Klein was often heard saying,

> [The federal research restrictions] are really extraordinarily burdensome and represent a radical intervention of government in the right of a family member to access for the future the best care for their child, their parent, or their spouse. . . . So [Bush's policy] is highly disconnected from the American tradition, the American experience.[41]

Through such statements, voters' attention was successfully redirected from the socioeconomic barriers that have prevented most people from accessing "the best care" for their loved ones (and well before Bush's stem cell restrictions), even to the point where a bill to explicitly facilitate greater healthcare coverage was voted down in the same election cycle. Likewise, it was a desire to *insulate* science rather than throw open the field's doors to the multitudes that drove proponents of stem cell research to create a new "right to research" in the state constitution.[42] To do so, they effectively appropriated the political framing of their opponents—specifically, the Catholic Church and other "family values" organizations—arguing that families had a fundamental right to seek new and effective medical care for their loved ones. The official slogan of the umbrella stem cell advocacy organization Americans for Cures is "Pro Families, Pro Cures." Enabled by their resource-rich campaign, the proponents' messaging saturated the media-driven debate and ensured their effective use of the state initiative process, California's "fourth branch of government," as a tool to protect controversial stem cell techniques.

Beyond Biomedical Consumption

The challenges associated with including a heterogeneous citizenry, or *demos*, in the development of controversial life sciences rest in part on the fact that people's bodies, or *bios*, are implicated in the process. The use of human cells—tissue necessary for reproduction (embryos) and heredity (gametes)—further complicates the usual model of developing science and technology, in which people are positioned as future consumers of biomedical goods. Instead, stem cell research employs human tissue in the *production* of biomedical goods, without fully conceding the political and economic impact of such utilization. For this reason, the quandary of "peopling" stem cell research involves both the *participation* of the *demos* and the *recruitment* of suitable *bios* as tissue donors and test subjects.

In our post–World War II context of science development, people who donate tissue for research are subjected to elaborate ethical protocols, while any rights or interests they may wish to assert in the growing tissue economy remain largely untenable, according to the apolitical framework of bioethics itself.[43] Human tissue is absolutely essential for stem cell research, so the challenges presented by a multitudinous demos asserting interests beyond access to biomedical goods potentially slow the development of biotechnology, as illustrated by the Oakland forum. Proponents of rapid scientific development have thus had to try to derive ways by which to include people as bios in ethical experimentation while holding the demos at bay.[44]

The double evocation of both of these words—that a heterogeneous citizenry (demos) needs to *demonstrate* its civic potential through direct action, boycotts, and petitions, and that people's bodies (bios) are shaped by complex personal and collective *biographies*—points us to the tensions that bubble beneath the surface of participatory rhetoric. This is not simply about the rhetorically slick packaging of an initiative but also about how the "black box" of participation is assembled and contested through competing claims to who the demos and bios of stem cell research are and should be. For Zach Hall, it was mainly about ensuring that tissue donors represented the genetic diversity of the populous, but for many at the Oakland forum it was about who would benefit from the work contracts and educational investments that grew out of the grant distribution process. After all, safeguarding the autonomy of California stem cell research would require building new facilities separate from those in which federal grant money would be used, among the many other tentacles of the initiative. Too often, this larger social context is overshadowed by a narrow focus on the speedy production of biomedical goods.

The assertion of rights and responsibilities narrowly tied to individuals' "fleshy, corporeal existence" is what has come to be known as biological citizenship.[45] For example, women who have a heightened genetic risk of breast cancer have organized around this shared biological experience to enhance their treatment options. Much of the analysis of biological citizenship tends to focus on social collectives "beyond the state,"[46] which also tend to include a more elite strata of people who have the resources and influence to organize around their shared interests. Within the pages that follow, however, we find greater ambivalence and at times pushback against this biological definition of citizenship than is evident in neighboring arenas.[47] Benefits from biological citizenship are not equally available to all social strata—potentially exacerbating inequalities, as it relies so heavily on a notion of depoliticized consumption and market optimism. The "biotechnical embrace"[48] that is welcomed by stem cell supporters who desperately seek breakthrough treatments may amount to a deadly constriction of opportunity for subordinate socioeconomic groups who have a different relationship to science, medicine, and the market. For people who are un- or underinsured, for example, advocating for novel medical goods when access to basic health screening and existing treatments is out of reach is in effect an invitation to a roulette game in which, it seems, their number is never called. So for many impoverished people, ethnoracial minorities, and people with disabilities, all of whom have been historically exploited or neglected by scientists and doctors, excitement over touted "scientific breakthroughs" may generate skepticism regarding the potential of regenerative medicine to address their well-being.

In bringing the resistance of subordinate social groups into view in a way that the literature on "biological citizenship" has tended not to do, I suggest that thinking "beyond the state" grows more problematic as many elite nonstate actors manipulate the state apparatus to protect their interests.[49] As sociologist Patricia Hill Collins explains, just as subordinate classes gained access to full political citizenship and laid claim to public entitlements, they were "increasingly abandoned by individuals and groups with power. . . . In this context, 'public' became reconfigured as anything of poor quality, marked by a lack of control and of privacy. The public sphere becomes a curiously confined yet visible location that increases the value of private services and [of] privacy itself."[50] Hill Collins's insights push us to recenter the state as a vital terrain of power and knowledge wherein state authority and oversight are experienced differently across social classes, while many in the pro–stem cell movement use a new state agency to obtain freedom from broader government oversight.[51]

Indeed, the question of the relative centrality that the state should have is one of the main points of contention among advocates on all sides. Proponents of stem cell research, while initially using the ballot initiative process to institutionalize this public science-making venture, actually structured the proposition so as to *limit* both the California and the federal governments' ability to interfere with the science. Opponents and critics of the initiative, for their part, have sought to enhance the authority of the state to intervene with respect to the initiative's implementation. In limiting the legislature's authority, the architects of the initiative also strategically constrained the ability of social groups to impact the implementation of Prop. 71. With the California state constitution now amended to support the "right to conduct stem cell research,"[52] those implementing the initiative were afforded a legal fortress from which to exclude their ideological and procedural critics.

While we find some evidence that people are attempting to push back against systemic exclusion,[53] it is important to note that we don't usually find fully formed collectives engaging with or opposing the initiative, but more tentative alliances, forged within a context of scientific uncertainty and social insecurity.[54] By examining intragroup debates about what actually constitutes "disability interests," "women's interests," and "ethnoracial minority interests," we necessarily find that collective responses to scientific initiatives are in flux.[55] Before a cohesive platform or slate of demands has been articulated as *the* feminist or *the* racial justice alternative to Big Science, we find more open-ended contests over the issue of stem cell research, usually with a greater variety of alternatives at play. But taking note of this instability is not meant to be purely an academic exercise in social deconstruction for deconstruction's sake. Rather, it is the first step in reconstructing a social architecture in which people actively struggle over both the outcome and the process of stem cell research, so that we can ultimately imagine better, more just, and more inclusive alternatives. Examining the process of peopling a people's science, along with all the exclusions and censorship of intragroup heterogeneity that often entails, will hopefully help us construct more socially robust futures.

People's Science, then, is not a romantic rendezvous with populist ideals, nor a resigned arranged marriage with politics as usual, but a somber yet hopeful blind date in which we search for clues about what lies beneath the slick packaging of "the people." Beneath the seeming win by the "Yes on 71" campaign, I take you beyond election politics to the bioconstitutional moments before and after that campaign, revealing episodic struggles over the use of human bodies and the rights owed to human beings that do not always become codified, precisely because many collectives lack the necessary power to influence the governing apparatus. This encourages us to seek out not only "ethical" formulae to manage science, which all too often preoccupy those involved in stem cell governance, but more politically inclusive and socially just practices that force us to grapple not just with what is ethically right but with *what* rights people are owed.

Investigating the Initiative

To investigate how social groups attempt to shape and intervene in the development of stem cell research in the midst of immense uncertainty about the fruits—bitter and sweet—of that research, I conducted a multisited ethnography of the California Stem Cell Research and Cures Initiative from August 2005 to August 2007, across three main sites: biomedical, regulatory, and civic. Through a formal affiliation with the state stem cell agency as part of its first cohort of "training fellows," I participated in meetings, conferences, retreats, academic courses, and legislative and legal hearings. The CIRM (California Institute for Regenerative Medicine) training program was created to draw young scientists in relevant scientific fields across the state into the emergent stem cell field. As the only social scientist, I was expected to research some aspect of the "ethical, legal, and social implications" (ELSI) of the field and required to take a four-month-long bioengineering course as well as an ELSI course; meet with other fellows in a monthly journal club for nine months; and provide regular updates of my research activities to the state agency over this entire period. During these activities, I was able to observe the way in which the agency's goals were clarified, as well as how socially dispersed actors (scientists, bureaucrats, lawyers, bioethicists, venture capitalists, politicians, patient advocates) coalesced around the shared goal of legitimizing the initiative in the face of two lawsuits and a proliferation of critical news stories.[56] I also analyzed Prop. 71 campaign finance documents, CIRM press releases, and transcripts of those committee meetings I was unable to attend, in order to compare the backstage and front-stage dynamics of implementing the initiative.

The legal battles that served to delay the agency's ability to access money from the sale of state bonds had the fortunate side effect of allowing for a more protracted postelection battle of *ideas*. So while CIRM was forced to justify its legitimacy in a court of law, CIRM officials also had to make their case in the court of public opinion. During this postelection interval (November 2004 to March 2007), CIRM was unable to begin the business of handing out grants to researchers and institutions, as its explicit mandate requires.[57] The institutional insecurity produced by the postelection lawsuits, in conjunction with media skepticism, allowed me to observe the struggle over legitimacy in a more elaborated form. It was within this window that a number of public policy organizations, watchdog groups, journalists, and academic critics became even more outspoken about the ethics of research, the lack of oversight and accountability for Prop. 71, and the conflicts of interest plaguing the new agency. I also trained my attention on those civic groups, advocates, and organizations that were actively contesting or defending the parameters of inclusion in the initiative. This included examining four groundbreaking forums, including the Oakland forum described earlier, and to a lesser extent analyzing virtual debates occurring on key advocates' blogs and websites.

During this intensely stressful period for CIRM, I sought to temper my exposure to the growing public critique of the stem cell agency, and of Big Science more generally, by witnessing the firsthand experiences of families whose chronically ill children could potentially be helped by one of the most efficacious, yet still risky, stem cell therapies available to date: umbilical cord blood transfusions. I interviewed patient families, physicians, and other medical staff who utilized, or provided medical services at, an urban teaching hospital, which I refer to as the Garvey Research Complex. Here we find a cord blood banking and stem cell transplant program and a regional sickle cell clinic, as well as a number of other clinical and research departments.[58] At Garvey I shadowed the head physician in the sickle cell clinic and also attended medical rounds, observed patients who came for clinic visits, and took note of discussions between physicians, nurses, and social workers that included references to the medical and social histories of patients. I was given access to cord blood program enrollees and observed the process by which families collect and store blood for possible use in a stem cell transplant. Outside of Garvey, I observed patients and health professionals in a number of settings: home visits with patient families; sickle cell community gatherings; community-based talks by scientists working on stem cell transplantation. In addition to on-site interviews and home visits with clinic patients, I conducted remote telephone interviews with families who had banked umbilical cord blood from an unaffected sibling at Garvey's tissue bank for use in a stem cell transfusion. In all, I spoke with a purposive sample of sixty-three people who were actively engaged in advocating for, implementing, or critiquing the initiative in the biomedical, regulatory, and civic arenas.

Possessed of a keener understanding of the experiences of people living with illnesses potentially treatable by stem cells, I grew necessarily more empathic toward the rhetoric of urgency and hope that pervades discussion of stem cell research in the public domain. But I also became more attuned to the disjunction between the official representatives of suffering (for example, researchers or disease organizations seeking grants) and the diverse meanings that illness has in the lives of those on whose behalf Prop. 71 was passed.

Throughout my fieldwork, I added materials to a mixed archive I was accumulating of documents and media: campaign commercials, news articles, advocacy and stem cell policy blogs, opinion and exit poll data, patient newsletters, and research consent forms. My purpose in doing so was to examine how the specific themes that emerged during participant observation and interviews related to the mundane bureaucratic infrastructure involved in stem cell research as well as the wider ideological debate surrounding it.

Sankofa Science in Action

In ancient Roman mythology, Janus is a god of beginnings and transitions, two-faced because he looks towards the past and the present at the same time. In the classic text *Science in Action* (1987) sociologist Bruno Latour mobilizes the Janus metaphor to distinguish between "ready made science" and "science in the making," which we can simplify further as the tension between what scientists say science *is* and what they actually *do* in the day-to-day production of scientific facts. "This is what makes the study of the past of technoscience so difficult and unrewarding," Latour writes. "You have to hang onto the words of the right face of Janus—now barely audible—and ignore the clamours of the left side."[59] When I was first introduced to the Janus metaphor, I was struck by how much the Roman figure resonated with the traditional Ghanaian (Akan) symbol of the Sankofa bird, which looks back over its shoulder to retrieve an egg perched on its back. Here too is a creature who is being pulled towards the past and the future, expressed by the Sankofa proverb, "*Se wo were fi na wosan kofa a yenkyi*," which means, "It is not wrong to return and take what you have forgotten." The implication is that learning from the past will help us build the future.[60]

But why all this talk of symbols, metaphors, and proverbs as we embark on a discussion of science and politics? Stem cell research is a forward-looking endeavor par excellence, promising a new kind of medicine that restores our ill bodies to their healthy state, and California's Prop. 71, we are told, is the vehicle that will move us swiftly to a future in which these chronic illnesses that plague our families and communities will be a thing of the past. It is an endeavor with built-in claims and aspirations about what lies ahead and what lies behind us—precisely the kind of thing for which both Janus and Sankofa were crafted to illuminate.

But unlike Latour, who follows scientists around the lab to see first-hand how they make facts by assembling the right kind of social networks with people and things, our foray into the world of stem cell research is much less about this traditional site of science-making, the lab. For that reason, I suggest we draw upon a new-old metaphor to help us conceptualize the bioconstitutional struggles that have emerged around this controversial field, to account for the fact that what *scientists* say and do is not the only, or even primary, site of tension and contradiction. The Sankofa bird, with its implicit challenge to the dominant cultural motif of Western knowledge-production, its integration rather than transcendence of history, and its incorporation of the egg—so central to stem cell research—provides an alternative symbolic framework to begin this inquiry. In contrast to a "people's science" that is fashioned

Janus

around the interests of the social elite, a Sankofa-inspired approach to science critically engages histories of domination and subordination in order to produce knowledge that is committed to not simply biomedical consumption (as the epitome of capitalist freedom) but an ongoing process (not end point) of social liberation.

Looking backward and forward through the idiom of Sankofa, we can attend to the fact that what may bring healing and longevity for some may threaten the rights and dignity of others, in part because of the historically mediated relationship different groups have with scientific experimentation and biomedicine. If we ignore this history, we are fashioning a future with in-built inequities.

Consider the struggle between medical and social models of disease as articulated by deaf activists who do not want to be "cured" by stem cell research. This pushes us to think critically about how the line separating diversity and defect is drawn. Only by digging deeper into this tension can we appreciate how the imagination of the liberal citizen finds its foil in the relations of care and dependence that are often pejoratively assigned to the lives of ill and disabled people. This includes even the way our "very language and imagery of citizenship" equates independence with the able-bodied: "We speak of upright and upstanding citi-

Sankofa Bird

zens, we stand to attention to the playing of the national anthem. The good citizen is embodied as male, white, active, fit and able, in complete contrast to the unvalued 'inactive' disabled Other."[61] Stem cell advocates, for example, not only express a compassionate desire to alleviate the suffering of those experiencing illnesses and impairments; many also link future cures to an imagined freedom that we all presumably seek from caring and paying for the ill and impaired in our midst.[62] In this way, regenerative medicine is imbued with the moral authority to unburden family members, society, and most crucially, the state from the burdens of care. The struggle between stem cell advocates and deaf cultural activists is *bioconstitutional*, as the biological and political statuses of bodies are contested as either tragically defective, and so requiring a cure, or triumphantly diverse, thus requiring protection of their self-determination.

But typically only one of these framings makes it on the official record. The "default public" of science is, not surprisingly, those who are cheering on innovation but having little tolerance for Sankofa-like questions that urge us to examine how our vision of the future impacts social groups differently. Despite important inroads on the part of the disability rights movement, stem cell proponents who make up the pro-cures movement are the legitimate constituency of this field. These patient advocates work to alleviate the suffering of impaired and diseased individuals and in so doing constitute the dominant demos of state-sponsored science—a demos that is conceived of as a collection of ill bodies seeking biomedical goods. This flies in the face of decades-long disability activism that sought, in part, to counter the prevailing approach to people experiencing illness or impairment as always needing to be fixed—an approach that pushes us to address the way that *social* disorder is perpetuated through the allocation of resources and normalization of stigma directed at vari-abled individuals. The monopolization of the public domain by patient advocates who do not tend to incorporate this social history into their vision of the future compels us to consider how scientific advances may lead, at once, to political retreats, in often unintended ways.

This uncertainty about how developments in stem cell science are impacting the social landscape—including the exercise of power, accessibility of biomedical goods, and cultural associations between disability and civic agency, to name but a few—is why it is so vital that we not only take historical precedents seriously but also develop "technologies of humility,"[63] which, I suggest, is not so much about submission as it is agility. The hubris that Latour links to "ready made science" is nowhere better exemplified than in Nobel Prize-winning geneticist James Watson's statement that if "we [scientists] don't play God, then who will?"[64] This, I submit, is a kind of false security. On the surface it seems to lend authority to new fields like stem cell research that are attempting to establish their legitimacy and actualize their promise. But the underside of security is captivity—fixing things in place at the very moment when we seek to transcend the bounds of medicine. In many advocates' hyped-up rhetoric we find a lack of agility in being able to deal with the pitfalls, the unintended consequences, and the differential impacts that innovation necessarily produces. In cultivating greater flexibility to deal with the unexpected—like the Sankofa bird twisted back upon itself— we need a more humble approach to science-making that takes the time to incorporate our many social histories into a just and equitable vision of the future.

CHAPTER ONE

LOCATING BIOLOGICAL CITIZENSHIP

> Stem cell advocacy is not a political movement. It is a consumer movement! If you ask people on the street if they support this, they do, not because it is a public health issue, but because it's a personal health issue.
>
> —*Bernard Siegel, stem cell advocate*[1]

> Our parks are closing. Our education budget is being slashed. Our infrastructure goes unrepaired. Cops are being laid off. Our university students' tuition is shooting through the roof. Kids are being thrown off Medicaid. . . . But the CIRM keeps borrowing from the impecunious to pay for its fat salaries and luxurious buildings.
>
> —*Wesley J. Smith, consumer advocate*[2]

"LOCATION, LOCATION, LOCATION!" is the enduring indicator of value with respect to prime real estate; but in a nongeographical sense, it serves to signal social worth as well. Where, on an individual level, one is "located" within crosscutting social hierarchies—for example, whether one is a Hollywood executive with a bevy of top medical specialists on speed dial versus a drugstore clerk who turns to the ER for medical care, and then only with the most unbearable maladies—can be a matter of life or death.

Our position in the social world gives us a particular vantage point with respect to everything from the new organic supermarket moving into our neighborhood to the latest scientific innovation that promises to regenerate our relative's stroke-induced paralysis. For the executive, the new supermarket is perhaps one more welcomed option; for the clerk, such neighborhood revitalization likely means her rent will increase, forcing her to move out. There is little wonder then that residents in a growing number of transitional neighborhoods throughout the country have attempted to protest the construction of Whole Foods Market stores as a visible symbol of their impending displacement, rightly inferring the inverse relationship between more healthful food options on their doorstep and their ability to keep up with rising property rates. In our enthusiasm for expanding healthful food options as part of serving the collective good, we neglect the larger social context in which goods are brought to market. Even analysts who might otherwise critically attend to these dynamics can be swept up in the promise of regeneration. Sociologist Loïc Wacquant points out this

> troublesome trend in recent studies of gentrification, whereby the takeover of working-class districts by middle- and upper-class residents and activities is increasingly presented wholesale as a collective good. . . . By focusing narrowly on the practices and aspirations of the gentrifiers through rose-tinted conceptual glasses, to the near-complete neglect of the fate of the occupants pushed aside and out by urban redevelopment, this scholarship parrots the reigning business and government rhetoric that equates the revamping of the neoliberal metropolis as the coming of a social eden of diversity, energy and opportunity.[3]

The broader social context is one in which the individualistic logic of free choice is rather costly for those who cannot afford all the upgrades taking place in the public domain. Returning to the issue of regeneration in the biomedical context, one young Filipino American man who lost part of his lung to tuberculosis observed that

> the promise of therapeutic treatments derived from stem cell research gives individuals like me a hope for normalcy. Yet, as an immigrant from a low-income family, I can't stop from cringing at the thought that the low-income and marginalized communities of the state still have no explicit guarantee of access to the promised "cures" of Prop. 71—much less to adequate health care in general.[4]

Another man born with cerebral palsy asked whether "as a Black, disabled activist living on SSI [Social Security Insurance], would this proposition reach my people and other people of color who are wheelchair users because of police brutality? . . . With $3 billion going toward this research, how much will go toward social programs, health care, and the run-down hospitals in our cities?"[5]

By contrast, in the epigraph to this chapter, stem cell advocate Bernard Siegel asserts that proponents of this research are part of a "consumer movement" for more and better choices in treating currently incurable illnesses—a "personal health" as opposed to public health issue. In isolation, who would object to the amelioration of sickness via more effective therapies? But despite numerous physical and symbolic attempts to erect walls, build gates, pave private ways, and create social closures so as to separate "us" from "them," our life chances and well-being are not simply "personal" but interconnected. We cannot afford to examine any campaign for public underwriting of stem cell research as a movement to produce biomedical goods without locating it within broader systems of power, inequality, and the collective good. The relationship between our social positions and the positions we take on the question of stem cell investment is reinforced by how much power we do or do not have to pull the levers of influence in response to our concerns and interests.[6] That is, the higher our position in the social landscape, the more the objective world (institutions, policies, laws, and so on) reflects what we hold most dear. So, despite occasional delays, the Whole Foods Market eventually moves into the neighborhood, and residents unable to afford the higher cost of living must eventually move out.

In one of the most organized community campaigns to first resist, then engage, the supermarket company, residents of Jamaica Plain, Massachusetts, joined together in a "Whose Foods? Campaign," asking the company to sign on to a Good Neighbor Agreement and donate 1 percent of its annual revenue from the local store to fund "local anti-displacement organizing . . . and the creation and/or preservation of local affordable housing" for the duration of the store's twenty-year lease. In demanding "a small slice of the pie," the Whose Foods? Campaign is akin to the efforts of those seeking to establish a mutual relationship between biotech companies that benefit from CIRM (California Institute for Regenerative Medicine) grants and California residents unable to afford future stem cell therapies.[7] By requiring that royalties be paid to the state, the exacerbation of existing inequities can be partially mitigated.

The Unfolding of Proposition 71

The political experiment that is Proposition 71, in which new technologies in the public domain provide the scaffolding for scientific experiments, is not simply encapsulated in the ballot measure in which citizens cast a vote for or against public investment in stem cell research. The experiment is actually ongoing, conducted in local and episodic[8] public engagement exercises that attempt to bring together the "right" kind of publics, as imagined by loyal enthusiasts of the science, to determine the ethical and procedural rights and wrongs of the initiative. Rather than contributing to the creation of some underlying social consensus, such engagement is better understood as a series of credibility struggles that are performative and eventful,[9] especially when the veneer of populism cracks in the face of the "wrong" kind of audience participation.

To identify the civic stakes involved in the California initiative, I take you behind the scenes of three forums in which the "right" public was painstakingly assembled and participation enacted. Then I examine a fourth case, in which the "wrong" kind of public was at the stem cell governing table, and I discuss the negative backlash that ensued revealing the fragility of such participatory arrangements. I draw upon Herbert Gottweis's examination of public participation in the European Union to suggest that California's effort to act on behalf of the common good is necessarily a treacherous undertaking when who counts as common and what counts as good are themselves contested.[10] In these participatory episodes, we see how advances in the life sciences are giving rise to new demands and new rights claims—though they are not wholly new, because of how social elites have come to conflate "what is good" with unfettered access to biomedical goods.

Using bioconstitutionalism as a framework by which to understand the relationship between biological and political experiments, California's novel "right to research," codified in Prop. 71, must be situated within a civic context that values particular kinds of publics that are, first and foremost, wholly committed to the swift, no-holds-barred advancement of stem cell research. Paradoxically, despite the populist packaging of the initiative, many stem cell enthusiasts view regulations that aim to ensure that future stem cell therapies are affordable as an attack on their personal right to access therapies as soon as possible, because of how such impositions may disincentivize the biotech industry from moving research from "bench to bedside."

Even so, despite what might seem like obvious class-based differences in the expectations people have with respect to scientific innovation, collectives that might give voice to these competing concerns do not come ready-made. Rather, in the bioconstitutional struggles to follow, we see how the people who argue on behalf of a stem cell consumer movement or social justice movement come together through a process of participatory fashioning, which occasionally breaks down. Even those civic spaces that are created, within which to critique or challenge the initiative, require work to generate sufficient interest in the "goods" of stem cell research; otherwise, why would we want fair access through redistribution policies, such as higher state royalties for biotech companies that use CIRM grants? All stakeholders—both avid stem cell supporters and antagonists—are defined through a process of knitting people's existing interests and concerns to the promise of this new field.

The point of this discussion is not to adjudicate the relative authenticity of participation, deciding when and where the "real" public is present and what the "real" interests of "the people" are. For many observers, the exclusion of undesired publics started when the proposition's architect, Robert Klein, crafted what California's Little Hoover Commission called an "insider's club" by writing in exemptions to the state's open meeting laws, guaranteeing that university administrators who were eligible for CIRM grants would also serve on the agency's governing board, among other procedural arrangements that could fuel conflicts of interest. While such ruses contribute to the overall context of organizational insulation, focusing on them places undue emphasis on procedural questions of transparency and openness rather than asking how particular publics and interests gain currency (or as social theorists call it, "hegemony") in the first place. How did the architects of the initiative create a populist veneer through actually *avoiding* (rather than institutionalizing) conflict with subordinate groups, in a context that is technically open but in practice closed?

The Politics of Proximity

On an organizational level, the question of where to locate the stem cell agency (CIRM) was one of the first crucial issues in the practical orchestration of civic participation. Choosing a location, both geographically and in the broader network of organizations, involves considerations of how best to communicate institutional identity, to cultivate legitimacy, to generate resources, and to align oneself with the right sort of publics. Whom or what an organization is close to shapes how it is perceived and what kinds of constituencies have access to it or are excluded. Whereas news reports typically cover medical breakthroughs by taking the public inside laboratories, interviewing scientists and reporting their novel scientific discoveries and techniques, the social domains that scientists themselves rely upon for material support (for example, grants) and symbolic resources (such as legitimacy) often remain hidden. These less visible venues, where nonscientists often join the debate, are a central feature of what social analysts call the "new government of life."[11]

If, as one patient advocate observed, California is "the stem cell state," then the question of where to locate the headquarters of the stem cell agency comes down to a matter of designating the "capital" of the initiative. At first glance this decision may seem to be a mundane bureaucratic exercise like any other, but as we will see, the debate over the location of the headquarters was infused with hopes and anxieties about access to stem cell research and the exercise of other forms of symbolic (status and prestige), social (networks), and economic capital. These other assets facilitate or block structures of oversight, webs of accountability, and economic and scientific exchanges in ways that can generate growth or bring about stagnation in a fledgling agency. Focusing on bricks and mortar, breaking ground, and building walls serves (especially for me as a sociologist, trained to think in terms of *social* construction) as a window through which to explore how power is made, exercised, and consolidated through seemingly "public participation." The material construction of the stem cell headquarters and the social construction of public participation in stem cell research go hand in hand.

Leading up to the final vote of the Independent Citizens' Oversight Committee in Fresno on May 6, 2005, which was in favor of San Francisco's bid to serve as headquarters of the stem cell agency, tensions had been growing in a series of public meetings held up and down the state. Candidate cities had incorporated all manner of enticement in their application packages in a quest to come out the favorite—everything from, in one account, "free upscale office space, furniture, utilities, business and recreational services, . . . parking, [and] security" to a grander array of perks such as "occasional private jet use."[12] Among fifteen cities originally in the running, Sacramento, San Diego, San Francisco, and Los Angeles made the short list, the last disqualified late in the process.[13]

In part, one might say that from the perspective of the candidate cities, the decision over the headquarters was an issue of *branding*: bidders used the reputation of their cities to assert distinction in one sphere or another. Thus, a vote for San Diego, since the city was ranked first in the country as a regional biotech hub,[14] was for a more *science-centric* approach to the initiative; a vote for Sacramento—the state capital, where any group seeking to influence the legislature has an office—was for a more *politics-centric* approach. A vote for San Francisco—a liberal, cosmopolitan refuge with the most international flights to Asia—was for a *global economics–centric* approach to the initiative. San Francisco sought first and foremost to position the new agency as an international player, on a par with other national stem cell initiatives and with sufficient autonomy vis-à-vis the regulatory state apparatus to fund the most cutting-edge science.

In practice, of course, all of the above qualities are important and also interconnected, so the decision was more about what the stem cell field should value *most*. The competing answers to this question were animated by the relative importance of research sophistication, traditional legislative governance, and global prowess in the agency's mission. The stem cell agency is now situated in an upscale office complex across from AT&T Park, where the San Francisco Giants play, reminding us how this early vying over siting the headquarters was part of a larger social field infused with existing power relations. In characterizing social action, social analysts regularly draw upon the metaphor of a *field*, wherein people compete over symbolic, social, and economic capital within predetermined parameters and according to prescribed rules.[15] In recent examinations of public participation, analysts also point to the way in which the rules of the game—"the way in which a problem or issue can and should be legitimately framed and publicly handled—are themselves the subject of political deliberation and struggle."[16]

In explaining the rationale behind the final decision, CIRM board member Jean Fontana explained how the decision was a reflection of the members' vision of the institute not as a place that "robotically process[ed] grant applications" (that is, not one that was science-centric) but as one "providing national or international leadership with respect to cutting-edge thinking about how to defeat these diseases."[17] Robert Klein, architect of Proposition 71 and then chair of the CIRM, explained that "This is to lead the world. It is to lead the nation, and those resources [referring to the free conference facilities provided by San Francisco] are vital to it."[18] He goes on to depict San Francisco as the most cosmopolitan of the options, whose facilities could house thousands of people "from all over the world"; a site so liberal and supportive of stem cell research that even top members of the religious community formally supported the city's bid. He envisioned the institute, based in San Francisco, as sufficiently open to the international stem cell community and sufficiently insulated against critics of the new agency. "Authors of the initiative, wary of government interference, made CIRM's funding virtually independent from state government . . .

California responded by creating a safe harbor, free of political influence, for scientists to conduct such research. The political climate has since reversed"[19] since the Obama administration loosened restrictions on the stem cell field.

Criticizing the final choice, then California state senator Deborah Ortiz spoke on behalf of Sacramento's bid, saying a vote for Sacramento was a vote for "the voice of the people of California." Basing the agency headquarters in the state capital, she argued, would encourage a "participatory process" whereby advocates could engage with the agency and "debate this great dream." Other Sacramento supporters touted the advantage of their site in terms of its home to "ethnic chambers of commerce" who had signed letters to support the city's bid. José Perez, publisher of *Latino Journal*, explained that the "Latino community in California is huge. It's like eleven million people, 33-plus percent. If we take a look at the governing body right here, the question is, does it look like that? . . . and interestingly where there's greatest isolation for Latinos is actually in the Bay Area. It's in San Francisco."[20] He urged the selection of Sacramento as a place symbolizing greater solidarity with the state's largest ethnoracial minority.

Notably, Robert Klein did not engage with the interests of racial-ethnic minorities and those arguing to prioritize access for interest groups who sought to "debate this great dream" throughout the implementation of the initiative. In fact, Klein implicitly structured the initiative, and successfully sought to influence the decision over the headquarters location, with the aim of creating distance *from* politicians and interest groups. California's new constitutional amendment, which Klein called a "legal sanctuary" in one interview, served as a shelter against those who might further politicize this already controversial science.[21] And for its part, the agency from its inception aligned its own mission with those interested in insulating the initiative from traditional interest-group politics (represented by Sacramento) in favor of greater international networking and competition (represented by San Francisco), because the latter promised to fast-track scientific breakthroughs and biomedical treatments. At least, this was the hope in locating the agency there.

As with other capital ventures, investment in an autonomous stem cell capital was a highly speculative one.[22] Crucial to the process, then, was the cultivation of stakeholders. The extensive process whereby the original fifteen cities mobilized local chambers of commerce, mayors, city councils, religious organizations, and all types of local advocacy groups and constituencies in support of each respective application is itself instructive. This labor of amassing widespread endorsements extended the agency's eventual reach far beyond what it would presumably have been if a simple top-down decision had been made about where to locate the headquarters. Proponents of each city bid cultivated interest in the headquarters contest among constituents who, prior to the process, were likely uninterested in the theretofore speculative science of stem cell research. But through this elaborate procedure of gathering stakeholders' written endorsements, an array of social, economic, and political concerns was knit to the headquarters siting decision, and a "people's science" began to be fabricated in the process.[23]

Despite the elaborate staging of public participation, there is no way to ignore that the CIRM chairman, Robert Klein, supported the San Francisco bid that eventually won. Even before the passage of Proposition 71, critics began pointing out Klein's conflicts of interest, suggesting that the qualifications for chairperson curiously overlapped with his professional background.[24] News report after news report called it "not very surprising" that Governor Schwarzenegger and three other elected officials (as per Prop. 71 rules) nominated Klein for the position, observing that the so-called stem cell czar had bankrolled and masterminded the initiative.[25] More sympathetic accounts noted how the diagnosis of his son with Type II diabetes, the death of his father due to heart disease, and his mother-in-law's diagnosis of Alzheimer's had all served to spur Klein to action. Through whichever lens one views the seeming inevitability of Klein's chairmanship, Parkinson's patient advocate Amy Comstock's praise that "throughout his career, and most recently as chairman of the Proposition 71 campaign, Mr. Klein has demonstrated an understanding of the politics of science"[26] would seem to hold.

Following Prop. 71's passage, not only did Klein's leadership come under increasing scrutiny, but so did the relative independence of the Independent Citizens' Oversight Committee (ICOC). More astute observers realized that the autonomy of the board was first and foremost in relation to state oversight. A number of local news outlets began picking up on the exceptionalism of the agency, noting that the text of Proposition 71 explicitly exempted the agency from open meeting laws applicable to all other state agencies, allowed it to self-monitor its members' conflicts of interest, and prevented the legislature from amending it for three years, and then only with a near-impossible two-thirds majority. One especially provocative editorial questioned the demographic composition of the ICOC, noting that most "are white, male, university officials. . . . the institute's oversight committee needs to be racially diverse, well represented by women and committed to the highest standards of ethical science and public accountability. The last thing we need is a committee of clones."[27]

Eventually, the heightened scrutiny forced Klein to resign from the presidency of the stem cell lobbying group Americans for Cures Foundation, in part because the group was advocating against a state bill (SB 1565) which would ensure that treatments developed using CIRM grants would be affordable for uninsured Californians. The organization urged its members to call their legislators to vote against the bill because, among other reasons, it would "discourage private industry from developing therapies and cures,"[28] and it then went on to publish a personal attack against the state legislator who sponsored the bill.[29] According to the organization, "The main problem with SB 1565 [is that it] introduces restrictions that will keep entrepreneurs, private investors and big companies from developing CIRM-based discoveries."[30]

Although Klein resigned as president of Americans for Cures, he remained on its board, while continuing in his position as chairman of CIRM. In another context, at a biotech industry conference on March 13, 2007, on a panel focused on corporate perspectives, Klein addressed the audience saying, "our primary mission is to advance therapies. . . . this whole issue of preferential pricing and state benefit has to be modulated against this primary mission. We have to consider how we're doing in terms of capitalist markets."[31] While I return to a fuller discussion of affordable healthcare in Chapter 6, for now it is important to note that while many stem cell enthusiasts voice support for affordable healthcare for all, CIRM leadership and their strident supporters, who approach this issue as constituting an all-or-nothing battle, view SB 1565 as a dangerous obstacle to the swift development of stem cell therapies.[32]

Those advocating for greater accountability and less insulation on the part of CIRM, routinely looking for connections between personal biography and stem cell policy, highlight Klein's background as a real estate mogul who made millions building affordable housing projects throughout California. As with the creation of Prop. 71 in response to Bush's stem cell restrictions, when Nixon ended public housing subsidies in the 1970s Klein and his business partner persuaded the legislature to "stage an end-run around the federal restrictions" by creating the California Housing Finance Agency using low-interest bonds.[33] At the same time, Klein came up with "new development-financing strategies in which market-rate units are included in subsidized projects to generate returns healthy enough to finance the whole deal."[34] In fact, the sale of one of Klein's housing projects helped finance the Prop. 71 campaign.[35] As with his potential conflict of interest as head of CIRM and president of Americans for Cures, Klein's work on housing finance required him to avoid using any money from the California Housing Finance Agency in his private real estate deals. A crucial difference between Klein's public housing advocacy and his stem cell advocacy, however, is that with the latter he "went straight to the voters in order to avoid having to deal with Sacramento legislative politics."[36] There is little wonder then that he did not support Sacramento's bid to serve as CIRM headquarters.

Drawing a direct parallel between Klein's two passions, urban redevelopment and cellular regeneration, chairman of Citizens for Responsible Government and libertarian blogger Wayne Lusvardi argued that "[s]tate-funded stem cell research is based on the same model as state-sponsored real-estate redevelopment":

> One of its key elements is the creation of the psychology of a race for new biotechnology and the elimination of blight. The psychology of

the redevelopment model and state funded stem cell research goes like this: If you don't build a new mall, or a stem cell research center, some other neighboring city, or state, will build it and other economies will thrive and yours will not. Just as in land redevelopment, the mission of stem cell research is to "eliminate blight." Biological blight is defined as "diseases resulting in sudden conspicuous wilting and dying of parts, especially young, growing tissues" caused by "a causative agent" that results in blight (e.g., cancer, heart disease, paralysis). In land redevelopment, "blight" is defined as "something that impairs growth or impedes progress and prosperity."[37]

In the history of urban redevelopment, efforts to revitalize the deteriorating inner city were typically carried out in the name of the working poor, but in practice it displaced families, concentrating them in smaller, more populated areas and thereby cutting them off from economic growth.[38] Some critics of Prop. 71 voice a parallel concern: that state investment in tackling "biological blight," without measures like SB 1565 ensuring that all Californians will be able to afford future treatments, promises to deepen existing health disparities. And in the decision to create both geographical and political distance between the stem cell agency and traditional legislative oversight, thereby protecting the intellectual property of the private biotech sector from overly demanding state royalty demands, the initiative implicitly catered to its "private" public and its "patient" public over other potential stakeholders. In what follows, I'll describe two other civic forums that followed on the heels of the "stem cell capital" decision, which demonstrate the competing values that animate this social field.

The Politics of Equity

On October 14, 2006, a group of social justice advocates, progressive policy analysts, and academics participated in a conference entitled "Toward Fair Cures: Integrating the Benefits of Diversity in the California Stem Cell Research Act." The stated purpose of the event was to "increase the understanding of the economic and medical potential of stem cell research among historically underserved minority communities and ensure that California's stem cell research efforts serve our state's diverse community."[39] The event was held at Children's Hospital Oakland Research Institute (CHORI), the former site of Merritt College—which the institute's president, Dr. Bertram Lubin, often reminded visitors was home to the Black Panther Party and their community health clinics.[40] In his introductory remarks, Lubin took note of the location's history as a way of conveying CHORI's ongoing public mission as well as the broad goals of the event.

As with the CIRM headquarters siting, this effort to increase the application of diversity to the stem cell initiative was not simply about bringing the right people to the table to discuss the issues. Rather, it required a small group of people who were already invested in the initiative to "people it" with stakeholders. Following Lubin's remarks, UC Berkeley chancellor Robert Birgeneau (UC Berkeley's Science, Technology, and Society Center was a cosponsor of the conference), CIRM president Zach Hall, and Greenlining Institute executive director John Gamboa set the stage for a discussion about social justice and stem cell science; Birgeneau strategically attempted to defuse an "us versus them" approach to the issues:

> One of the members of the ICOC who is a president of a biotech company said in the middle of the [ICOC] proceedings that "diversity underlies everything that we do." So there is no [question] that the stem cell oversight committee understands the importance of issues we're going to talk about today. And overwhelmingly members of the ICOC are committed to these ideals. . . . It's clear we need to address the issues of ethnic and economic diversity of our state in the implementation of Proposition 71 . . . to determine funding for scientific research priorities to benefit the most needy in our society and, finally, we need to make sure the scientists and entrepreneurs in this field represent the state's diversity.[41]

But if the ICOC was already "overwhelmingly. . . . committed to these ideals," why was there a need for such a conference? Indeed, among those gathered, there were competing understandings about the relative commitment of the agency to diverse inclusion as well as to what that might mean in practice. Although Birgeneau offered a staunch public endorsement of the Toward Fair Cures mission, this was only after he engaged in some hands-on, behind-the-scenes reframing of the agenda as a prerequisite for his support for the conference. One of the main conference organizers I talked with following the event explained that Chancellor Birgeneau required that organizers change the original subtitle of the conference from "Addressing the *Lack of Diversity* in Stem Cell Research" to "*Integrating the Benefits* of Diversity in the California Stem Cell Research Act" (my emphasis). As a major sponsor of the conference, the chancellor's office was adamant that the conference not be perceived as confrontational. While the chancellor's public comments express a commitment to "incorporating community participation and input in the governance of stem cell research in the very beginning,"[42] in practice the revised framing undermines an assertion of minorities' "rights" to be purposefully included in the initiative in the way that it sidesteps historical exclusion and subordination (that is, "lack of diversity").

In marked contrast, the Greenlining Institute's John Gamboa spoke about diverse inclusion as a political right accorded to California's ethnoracial minorities:

> Eighty percent today of all the households in California and for the next decade are households of color. Those are the households that are going to be paying the price of the bonds for the stem cell research. And those are the households that should benefit in all aspects of the benefits of this groundbreaking research. Those households should share in the jobs. They should share in the contracts. They should share in the research on it. The illnesses that affect that population are also included in the research of this stem cell movement.[43]

Joseph Tayag, then a health policy program manager at the Greenlining Institute, elaborated on this point, proposing that historically marginalized populations need to "be taken *off* the table (i.e., in clinical trials) and brought *to* the table."[44] This juxtaposition captures the tension between serving people's *needs* (when they are *on* the table) and engaging people's *interests* (when they are *at* the table), paternalism being embedded in the former and political empowerment associated with the latter.

But it is worth bearing in mind that engaging with the interests of stakeholders is no straightforward process, as we observed in the headquarters siting debate. Rather, over an eight-month period, Tayag conducted focus groups, made presentations and phone calls, and sent e-mails to convince the organizations in the Greenlining coalition to attend the Toward Fair Cures conference. He experienced more than a little hesitation, especially among religious organizations who expressed wariness about the use of embryos in stem cell research. As with the managing of the various city headquarters bids, it required that Tayag actively people this "people's science" with stakeholders who were, in some cases, reluctant to hold a stake.

Much of the Toward Fair Cures conference was geared toward addressing two main questions: How could the stem cell arena benefit from including historically marginalized groups? And how would equity in the stem cell arena relate to broader questions of justice? The first issue was addressed to those we might call "stem cell insiders," or those already convinced of this scientific field's promise. The second was addressed to stem cell outsiders—those who had yet to be convinced that they should care about this initiative. The perspective of the latter group was expressed by Sujatha Jesudason, policy analyst for the Center for Genetics and Society, who observed that "all technology has power relations embedded in [it], [is] developed to benefit specific populations and [is] made available to specific populations."[45] She went on to offer a telling observation: "[I]f we were to go to minority communities and women's communities and ask them how to spend $3–6 billion, it's unlikely that they would say, 'On stem cell research.'" Her larger point was that "however diverse participants at the conference appear, it was still an insulated space that remains disconnected to broader platforms for social justice."[46] But, again, insulation from interest groups and advocates who might slow the speedy pursuit of stem cell treatments was precisely what the architects of the initiative had in mind.

Selling It

The "right" kind of public, those who wholeheartedly endorsed the pursuit of cures by any means necessary, and to whom the stem cell agency offered an open, even revolving, door, was exemplified by Americans for Cures. With the stem cell agency chairman Robert Klein as its former president, and serving as an umbrella group for a long list of patient advocacy organizations that were mobilized first in the struggle for passage of Prop. 71 and thereafter for its protection from the likes of SB 1565 and other "assaults," the organization considered itself to be on the front lines of the stem cell battle. On April 12–13, 2008, it hosted a "State of Stem Cell Advocacy" conference at the University of California San Francisco's Mission Bay Campus Community Center to regroup, share best practices, and bolster supporters from multistate legislative actions akin to Prop. 71.

One of the first speakers, executive director of the Parkinson's Action Network, Amy Comstock, asked rhetorically, "What are we advocating for? Better treatments, cures, and research opportunities without the interference of politicians." Bernard Siegel, founder and executive director of the Genetics Policy Institute, compared foes of stem cell research "to people throwing thumb tacks on the road as an ambulance is passing by." Remaining speakers offered a number of strategies for advocates to use in garnering support for their cause. Communications consultant Steve Allen urged participants to "persuade, don't educate," adding that "you don't want people to think, believe me."[47] Emphasizing the importance of storytelling, he explained how it was important not simply to focus on one's own experience but to "talk about all the people that you represent."[48]

Conversations during breaks typically started with people making some mention of their "disease affiliation." After hesitating a few times when the pertinent question was posed to me, and noting the barely hidden suspicion on the faces of other participants—as if to say, "Then what other motive brings you here?"—I soon found myself noting, to others' reassurance, my work with sickle cell patients. In Steve Allen's closing comments, he cited Human Genome Project director Francis Collins as observing that "everyone has five genetic predispositions that if activated would give us some type of chronic condition"; accordingly, Allen noted, stem cell research is "relevant to everyone." If such predispositions bind us in an imagined community of potential illness, then refusing to express solidarity in the pursuit of potential cures was a kind of civic defection based in the ignorance of one's own inherent biological defects. Or so seemed to be the affective context of the conference, glimpses of which I caught in the suspicious looks generated by my delayed response to the question of disease affiliation.

In a candid afternoon discussion of the behind-the-scenes process of constructing effective framing and messaging, Paul Mandabach, Proposition 71 campaign director, remarked that "although California was in the biggest budget deficit in its history and bonds had never been used for medical research," the ballot measure had passed. He described the process of filming dozens of disease victims and scientists for campaign commercials: his team eventually selected Irving Weissman (cancer researcher at Stanford University); twin brothers, one with cerebral palsy and one without; and two Michael J. Fox ads. Opinion research consultant Richard Maullin noted that the campaign had surveyed people about "what they understood about stem cells already," and eight out of ten people related to the diseases and conditions that stem cell research (SCR) promised to treat—which led to a "cures" framing of the value of SCR. Dozens of prototype commercials were written, some of which were translated into spots; then, in the course of a "rigorous testing procedure," audiences were shown candidate spots to see which ones scored the best and were rated highest in terms of credibility, persuasiveness, and interest, among other factors.

In the Q&A following this panel on "messaging and polling," Klein also explained how the name of the Independent Citizens' Oversight Committee had been chosen: after research showed that people were concerned about accountability, the word "oversight" was included. Working with a team of five lawyers and through a careful process of revision, Klein had drafted over two hundred versions of the ballot measure during an eight-month process. Thus early on it was decided that the words "embryo" and "human cloning" should be eliminated in favor of the less controversial "blastocyst" and "therapeutic cloning."

A final way in which the initiative had sold its populist image was via the representative structure of its governing board. The ten diseases that were granted seats on the board, so to speak, are in no way underfunded or "orphan" diseases neglected by the pharmaceutical industry. Rather, they are all "Big Pharma" diseases that are already the subjects of highly capitalized foci of therapeutic development. As the communications director for the University of California San Francisco's HIV/AIDS program, Jeff Sheehy, observed to me in his role as the stem cell agency's HIV/AIDS disease advocate at the time, his participation on the stem cell board was rather ironic, given that his (disease) community is not really invested in stem cell cures:

> Sheehy: For a lot of us [in the HIV/AIDS community], we're not really thinking "cure." . . . Are we talking about something that's going to have any impact on the forty million people in this world living with this disease? . . . So we're not thinking about going on and trying this hypothetical cure, something that's going to be expensive, that's only going to be available to a few rich people. I mean, we're a community that spans the globe . . . none of us are really thinking, "Save me." We're all on the *Titanic* together, and there are no life boats!
>
> RB: So how would you compare your decades of experience doing HIV/AIDS advocacy with [the work of] those advocating for stem cell research?
>
> Sheehy: [Stem cell research advocacy] seems like a really top-down movement. . . . I'm the only person at the [committee] table making less than $100,000 a year . . .
>
> RB: Do you know how the board got structured?
>
> Sheehy: It's all [Robert] Klein . . .

RB: Do you know how the ten Independent Citizens' Oversight Committee disease seats were chosen?

Sheehy: I don't. But I cannot help but believe that there's a political calculus that went into it. I mean, I can't say for sure because I wasn't involved . . . [but] I mean, why were they calling me? Why were you trying to get HIV people to sign up? Because that's a very powerful, very active community. So it's obviously a constituency they were hoping would take them to . . . that 50 percent [of the popular vote] or up. It certainly wasn't [a drum] we were beating—we weren't demanding stem cell research. We were demanding drugs for Africa [laughs ironically]. We were demanding funding [for] the Global Fund. So that's where our heads were at. . . . We were like, fully fund the AIDS drug assistance program, so people in South Carolina don't die because they don't have access to the medications because they're not poor enough for Medicaid or because Medicaid is not funded in their state to pay for their drugs, you know. . . . That's where we were. We have drugs that can stop people from dying. So we were not concerned with the . . . promises of stem cell research.[49]

In short, the representation of ailments with powerful constituencies on the governing board was directly tied to the larger populist packaging of the initiative, even when the particular conditions themselves were already receiving investment money for research or when stem cell research was not likely to play a significant role in their alleviation in the foreseeable future.

Honing the right messaging and including powerful stakeholders were facilitated by an unprecedented budget for a state campaign, and even more so for a ballot measure: according to my tabulations of campaign finances, the "No on 71" campaign budget ($624,000) was less than 3 percent of the "Yes on 71" budget ($34,000,000).[50] The latter was able to harness all the levers of symbolic power, including television and radio ($14,000,000), drawing upon the expertise of consultants

($1,300,000) and saturating public discourse with its pro-cures message through petitions ($2,700,000) and other campaign literature and paraphernalia amounting to $3,000,000.[51] The cost of just one of the "Yes on 71" television ads exceeded the entire combined campaign budget of the "No on 71" organizations.

Then, paralleling the efforts of Toward Fair Cures organizers to garner support from their coalition members, the moderator for the final session of the State of Stem Cell Advocacy conference asked panelists whether they "had any problems with your organization supporting Prop. 71?" Siri Vaeth Dunn, mother of a thirteen-year-old daughter with cystic fibrosis who is "hooked up to machines," said that she "was stunned [that] at the CF Foundation board meeting [organizational support for Prop.71] was not a slam-dunk."[52] She described the foundation as a "fund-raising powerhouse" and noted that there was a fear among board members of losing donors "because of a knee-jerk reaction against embryo research."[53] Jeannie Fontana, executive director for patient advocacy at the Burnham Institute for Medical Research, noted a similar struggle with the ALS Association board of trustees: "The majority of trustees thought that maybe stem cells might be beneficial, but they did not want to risk going against the President [Bush] and the NIH [by supporting Prop.71], which might limit future ALS funding."[54] She noted that at an emotionally charged meeting, there was a neck-and-neck vote in which the organization decided to support Prop. 71 even though there were some very religious board members who voted no.

People's Violin

What happens when someone attempts to broaden the stem cell agency's accountability from the inside out, beyond patients who are already devoted to the goal of stem cell cures? Before her resignation as the diabetes disease advocate on the ICOC, Phyllis Preciado, a physician who served a predominantly working-class immigrant population in California's Central Valley, had gained the reputation of being, as it were, the "people's violin" insofar as she often recounted the hardships of her constituency. I first learned about Preciado in a conversation with *San Francisco Chronicle* science reporter Carl Hall[55]:

> Hall: She seemed to sort of have that one-string violin. You know, "What about Fresno? What about the Central Valley poor people?" . . . It was kind of an eyeball-rolling moment the few times she brought it up [at the board meetings]. I'm not sure I'm being fair to her. She may have had a point: "What about these people with diabetes in Bakersfield?"
>
> RB: Do you recall how her comments compared to the Parkinson's person or the heart disease person and how they would advocate for their [disease constituency]? Was she more vocal?
>
> Hall: I kind of thought so, perhaps; I don't want to sound unfair. I think she was sincere and everything; I mean, she is a medical doctor, I'm not. So, it's not like she is a stupid person; I never thought that. I just thought that her timing was a little off. And yeah, I mean, there are people who, [when] you give a citizen a microphone. . . . These social concerns become more of a show, then become a diversion or an annoyance. And then eyeballs start to roll a bit again. And people are like, "Okay, here comes the speech about the Fresno diabetics again."

Even for Hall, whose reporting on the ICOC and on Robert Klein was critical on a number of counts, Phyllis Preciado's references to Fresno could be interpreted as a "diversion" and "annoyance." Other disease advocates who spoke on behalf of their constituencies rarely if ever did so with reference to class or race demographics but rather through a strictly medical lens; whereas Preciado spoke of her desire to

> bring it back down to the people. So when we talked about grants and all those things, I would talk about, "Well, isn't it important also to spend some time educating our community about what is Prop. 71—*what is stem cell research*? And isn't it important to hold meetings where we actually educate our communities about . . . stem cell research?" So I became a broken record.[56]

Like the organizers of the Toward Fair Cures conference, Preciado was
attentive to the fact that those on whose behalf she served were not currently stakeholders in the process. If the board was to genuinely serve
them, she felt it needed to invest in educating the community. When I
asked her how her request to invest in a community education program
was received, she responded that while some board members may have
been supportive, "it takes time and money to do that. And their focus
with [respect to] the money is to, you know, do the research." In an
increasingly impassioned tone, she explained that

> [t]heir response to me was always, "Oh, let the newspaper [educate
> the community about the stem cell initiative]"—you know, which is
> such bull. If you look at all the monies we put into research . . . how
> much of that really, truly reaches our people? And usually the people
> who are benefiting are white—usually. So the disparities amongst
> the different ethnicities—it just enrages me that we're allowing it.
> And, you know, all these monies that are going into this program—
> how many brown and black people do you see really benefiting? . . .
> Don't get me started! I'm so upset by this.

Preciado described her work "in the trenches" when she oversaw a "care
for the poor program," with patients who were typically uninsured and
undocumented. As she recounted: "And so here I am trying to use my
voice here, and it gets misinterpreted because of my being aggressive,
which in a man would be assertive. Because I'm a female, I'm Latina,
and it's a battle. It's a battle that I fight. . . . They hire you, but they want
you to keep your mouth shut about other things." When I asked her if
she could recall any specific conversations or moments that illustrate
her "tokenization" as one of the only racial-ethnic minority board members, who, she believed, were not expected to actively try to shape the
ICOC agenda, she described a particular meeting in Fresno at which
one of the stem cell agency staff members scolded her for not showing
up at a previous meeting that finally included an explicit discussion of
community education. Responding to the contradiction between board
and staff members' simultaneous exasperation and expectation with her

community advocacy, this was the point at which Preciado made up her mind to resign. She explained to me, "That's not my job. My job is to bring it out. But [the stem cell agency's] job is to do it."[57]

Fed up with rubber-stamping the board's decisions without being able to substantively shape the goals and resources of the agency, she was frustrated with how little had come from her participation: "What have they done? They've done absolutely nothing. I would always say, 'Why don't we have educational forums? Why don't we go into the schools? Why don't we go to where the people are? Why don't we go to the community clinics?' Do you think that they're gonna give any money to community clinics? No." Much of the impatience Robert Klein and other board members appear to have had with Phyllis Preciado seems to have stemmed from her having introduced broader socioeconomic factors into the board's deliberations, thereby refusing to advocate solely through the lens of "her disease," diabetes.

Scholars describe a novel form of claims-making based on "biological citizenship,"[58] which is tied to advances in biomedicine and as a result of which citizens make demands upon the state for biomedical goods in a distinctly apolitical, market-oriented context. In that context, the "right to consume" even controversial therapies requires protection from what are described as "interest-group politics," including social justice concerns that might slow or thwart the quest for cures. Reciprocally, these biological citizens are expected to take responsibility for their own health as a matter of civic duty. But as legal scholar Dorothy Roberts points out in *Fatal Invention*, while much that passes for civic empowerment is more akin to consumerism, "it is a new form of citizenship that threatens to replace active, collective engagement to create a better society with providing information to the biotech industry and consuming its goods and services."[59]

In Phyllis Preciado's experience as a diabetes advocate seeking to link the medical plight of her constituency with their socioeconomic marginalization, we see the potential for the increasing power of biological citizenship to displace *social* citizenship claims. Indeed, biological citizenship is more or less available and appealing to different social strata insofar as it rests heavily on a notion of depoliticized medical consumption. Through its "participatory" structure, the California stem cell agency privileges disease advocates who expressly assert "rights" as a matter of consumer freedom, which goes hand in hand with the "right to research" codified in the California constitution through Prop. 71. However, both "rights" place few demands on the state to ensure that all of its citizens have access to the fruits of a $3 billion public investment. Stem cell research is part of a new "government of life,"[60] wherein people increasingly make claims upon the state that are related directly to their biological well-being and health in an effort to extend and enhance their biological lives. Those, like Phyllis Preciado, who might resist a strictly biologized and depoliticized rendering of their grievances and interests struggle to link their constituency's medical plight, and related biological citizenship, with social and political citizenship.

Those who question the forward march of deregulated research and market-based biomedicine are too readily categorized as enemies of scientific progress. Thus, bioethicist and SRC proponent Arthur Caplan portrays SCR opponents as a "bizarre alliance of antiabortion religious zealots and technophobic neoconservatives along with a smattering of scientifically befuddled anti-biotech progressives [who are] pushing hard to ensure that the Senate accords more moral concern to cloned embryos in dishes than it does to kids who can't walk and grandmothers who can't hold a fork or breathe."[61] But when we disaggregate Caplan's naysayers and attend to the different ways in which people are attempting to stretch the parameters of participation in this field, we observe that the rules by which the new "government of life" is being constructed serve a rather narrow "biomedical good." In bringing subordinate social groups into view in a way that patient advocates and the literature on biological citizenship do not tend to do, I suggest that the political insulation that Prop. 71 exemplifies is deeply problematic. Insofar as biological citizenship claims presume an autonomous individual working on his or her body in a more or less private arena free of state regulation, we need to give more concerted attention to how socioeconomic privilege necessarily informs this consumptive framework.

The ICOC's attention to the bioethics of stem cell research at the same time that it ignores the sociopolitical dimensions of this investment is part of a larger shortcoming involving the way bioethics ignores "social issues from being recognized or treated as ethical."[62] The institutional posture toward subordinate social groups is one that conceives of "difference" as something to be either celebrated (as in "integrating diversity") or suppressed because it could fuel distrust on the part of those groups. These approaches are two sides of a discourse inherent in American bioethics[63] in which sociocultural group difference, rather than the stratifying practices of powerful institutions, remains the focus of intervention.[64]

Phyllis Preciado, like many at the Toward Fair Cures conference, attempted to highlight this provincialism, as when microbiologist and legal scholar Pilar Ossorio asked for more focus on the "trustworthiness of institutions" rather than on the distrust of minority patients. But Preciado, as someone seeking to raise the possibility of engaging subordinate classes as interested parties in the implementation of Prop. 71, was poorly received: ignored when not simply derided. I suggest here that CIRM and others who give political nods to minority health needs without intending to redistribute the state's resources so as to address those needs fail to appreciate that "issues of redistribution are inseparable from questions of *dignitas*."[65] To celebrate diversity without engaging the broader concerns of subordinate social groups is invariably in the best interests only of the "inclusive" institution involved, and maybe of those token "diversity entrepreneurs" who are willing to rubber-stamp the institution's agenda without vigorous engagement and critique.

The Parameters of Debate

In the language of the sociology of science, we might be tempted to conclude that "cellular and civic potentiality are coemerging": that the large investment in stem cell research is animating new forms of activism and resistance. But in fact, I offer evidence that investment in cellular potentiality can have an inverse relationship to civic debate, which gets prematurely closed off in the singular pursuit of biomedical goods.

While I empathize with the siege mentality of many stem cell supporters who have had to look over their shoulders in fear of attacks from those who oppose research on religious grounds, this defensiveness has turned into an *offensive* against any and all proposals and critiques that are perceived as slowing the onward march of science. Yet for the sake of the science itself, this narrow view of what is in the best interest of the common good must be reopened for consideration.

Indeed, the price of the stem cell movement's insulation has become increasingly costly as the stem cell agency attempts to convince the public of its relevance as the initial ten-year funding cycle comes to a close. In an eighty-eight-page report, California's Little Hoover Commission suggested that Proposition 71 was already "a relic of another era" in the wake of Obama's election and his reversal of stem cell restrictions—which also raised questions, it said, "of whether the current level of [the state stem cell agency's] insulation still is needed."[66] The defensiveness of the agency, it said, is not relevant in the newly supportive research context. Even so, consummate stem cell advocate and vice president of public policy at Americans for Cures, Don Reed, expressed his deep frustration with and opposition to the commission's recommendations to restructure the agency. "I know nothing about real estate, except I think houses cost too much," he wrote. "So, based on the fact that I have a complaint, (justified or not) should I be put in charge of making a new law on how to run the housing market?"[67] Reed is a former teacher with a knack for drawing illustrative parallels; his query brings us back to "location, location, location" and the question of what kinds of participants can and should influence the direction of a state investment in science that is of potential importance to a socially stratified public.

Drawing upon Reed's housing analogy, what if his and so many millions of Americans' concerns about the housing market had actually been prioritized in the regulation of the banking industry in the years leading up to the 2008 economic collapse? Would not Americans' worries about affordability have tempered the blind pursuit of profits—akin to the quest for cures? Indeed, the trickle-down model is no more likely to apply in a biomedical context than in an economic one.

Of course, we can agree with Reed that technical expertise is necessary for sound decision-making: "[N]ot everyone is equally well positioned to formulate meaningful questions about science in a given policy context," as Sheila Jasanoff writes. But even she argues that technical expertise is not the only, or perhaps even the primary, requisite for participation in such ethically fraught domains as stem cell research. Consider the example of a behavioral geneticist who "may be far less capable of assessing the health effects of silicone gel breast implants than the scientifically untrained woman who actually wears them in her body."[68] From Don Reed's perspective, such a woman should not be in charge of making a new law to regulate the plastic surgery industry. And, of course, no *one* person should be entrusted with devising regulations for an industry that impacts so many different people. But this applies as well to the partial perspective of a geneticist or a surgeon who may lack important *experiential* expertise: robust decision-making requires a range of knowledge, including that derived from the experiences of those who may be helped *or* harmed by a given innovation.

I suspect that considering how much Reed and other stem cell advocates participate in the proceedings of the stem cell agency and share their own experiences caring for ill or impaired relatives at every opportunity for "public comment," their objection is not with participation in the abstract. Indeed, they praise the large size of the twenty-nine-member board, because they support the idea that a diversity of perspectives, as they define it, is required in such a weighty undertaking.[69] Rather, they oppose participation by those who might slow or stop the rapid development of cures. Given the heated opposition of those who want to completely halt stem cell research, such a litmus test for participation may be understandable. After all, those who oppose the field outright are looking not to participate in stem cell research decisions but to shut down SCR altogether. But in policing the gates of this social field in order to bar its most avid opponents, stem cell "true believers"[70] suppress thoughtful deliberation about how state investment in the field may impact people differently depending on their location in our social world.

CHAPTER TWO

WHOSE BODY POLITIC?

We consider such research an appalling waste of money. . . . Do they really think we are damaged and deficient people who are simply a biological mistake?

—*Paddy Ladd, Deaf activist*[1]

From this perspective seeing the bright side of being handicapped is like praising the virtues of extreme poverty. To be sure there are many individuals who rise out of its inherently degrading states. But we perhaps most realistically should see it as the major origin of asocial behavior.

—*James Watson, geneticist and Nobel Laureate*[2]

ON MARCH 12, 2008, the California Institute for Regenerative Medicine hosted one in a series of research spotlights on various conditions that are potentially treatable using stem cell therapies. On this particular day, at the Crest Theater in Sacramento, attendees gathered for a "Spotlight on Deafness," featuring Dr. Ebenezer Yamoah from the University of California–Davis School of Medicine. California Institute for Regenerative Medicine (CIRM) board member Dr. Claire Pomeroy introduced Yamoah, who was awarded a seed grant from the California stem cell agency in its first round of research grants, "with the goal of using stem cells to restore inner ear cells . . . the ones used to hear sounds and appreciate speech." In his presentation, Yamoah spoke with infectious enthusiasm about a technique his team was developing to regenerate tiny hair cells that are crucial for hearing. Unlike the cochlear implant widely used today, Yamoah's biological implant would preserve the structure and function of the sensory cells of the inner ear. In his words, "The ultimate goal is to make sure we can actually grow these cells, inject them into the inner ear, and hope they will incorporate themselves into the temporal bone and into sites where hair cells are lost."[3]

Pointing to one of his slides, he explained that "10 percent of the population has hearing loss caused by genetic defects, infections (meningitis and rubella), drug-induced, acoustic trauma, or age-induced" causes. As is typical at many of CIRM's public presentations, Dr. Yamoah embedded a deeply moral imperative to invest in stem cell cures throughout his technical explanations, noting that hearing loss has both a "psychological and [a] financial impact" and then describing the *social* implications of a deaf cure, invoking these words of Helen Keller's: "[B]lindness cuts us off from things, but deafness cuts us off from people." Yamoah continued by saying, "Human intercourse is truly defined by our ability to communicate through sound."[4] Through such seemingly well-meaning sentiments as this, Yamoah and others investing in a cure for deafness disregard the thriving Deaf world, in which people are not "cut off" from people but rather commune with each other through their shared experience of deafness. The value placed by the Deaf community on social intercourse via signing is so high that deaf parents have used in vitro fertilization to ensure that their children would be born deaf and thus more fully integrated into their social world.[5] So some members of the signing community contest the underlying assumptions of Yamoah and other stem cell researchers. "Scientists are patronizing the deaf by assuming they need 'curing.' . . . It's hearing people who have a problem with hearing loss, not deaf people."[6] For Lori Fuller and others active in contesting stem cell research, deafness is not impairment but an *identity*, and a marginalized one at that, requiring protection, recognition, and cultivation.

CIRM's "Spotlight on Deafness" thus not only ignores the self-definition of many in the Deaf world but may also pose a serious threat by making available very few outlets for debate or challenge. The stem cell agency's disease "spotlights" occur within highly circumscribed parameters of participation in which researchers and patient advocates define a range of conditions from deafness to diabetes within a strict medical model of disease. This medicalized approach reinforces dominant norms that place a premium on the "species-typical" body,[7] assuming the inherent desirability of hearing, walking, and seeing, for example. The medicalization of such functions also blurs the distinction between people suffering from painful and debilitating illness and those with sensory and mobility conditions that deviate from the norm. In their efforts to reconstitute the human body in novel (but normative) ways, proponents of stem cell research potentially marginalize those who do not conceive of their bodily condition as a defect but rather as part of human diversity. For this reason, it is important to distinguish between the well-meaning motivations of stem cell proponents and the ill effects of their advocacy. Consider three examples of how the meaning and experience of disability are contested.

Deaf Crusaders and Scientific Causalities

"Deaf Group Attacks Stem Cell Researcher for His Work on the Cure for Deafness." So announced a news headline on the popular blog of Deaf activist Mike X, who goes on to report that

> [a]rrests were made against a small group of Deaf people who attacked a stem cell researcher for his work on finding a cure for hearing loss. The attack took place at the researcher's home where he sustained multiple injuries. He was treated at a nearby hospital and released. The attack occurred while he was home alone. Graffiti was spray painted on the living room walls and elsewhere in the victim's house with words like "Audism is wrong" and "Deaf babies do not need your help," along with several epithets. The group, based on the graffiti, named themselves . . . "Deaf Crusaders."[8]

It turns out that Deaf Crusaders, and their alleged attack, were both fictitious—a provocation on the part of Mike X to generate discussion among his readers about the impact and meaning of stem cell research for the Deaf community. Those who responded point to precedents in other areas of science (attacks on researchers who experiment on animals, for example) and growing outrage by many in the Deaf world who view research seeking a "cure" for deafness as a form of eugenics. Although the ensuing conversation among Mike X and his readers did not express complete solidarity with the motivations or tactics of the Crusaders, neither did most of the participants align themselves with stem cell research and its pro-cures agenda. Amidst their ambivalence, however, respondents did tend to agree that a real confrontation between disability rights activists and stem cell supporters was just a matter of time.

Prognostications aside, it is interesting to consider how the fictional conflict scenario reproduced above contrasts with the relative dormancy of the struggle between, on the one side, researchers and patient advocates who are part of a growing stem cell movement,[9] and on the other, disability activists who view such research as an attack upon their right to exist and thrive as differently abled individuals and communities. This conflict bubbles beneath the surface of the prepackaged "stem cell debate," with few channels for disability-based critics to contest the norms and logics of regenerative medicine. It forces us to look directly into (appropriating Latour) the *Janus-face of technoscientific innovation*: What may bring healing and longevity for some may threaten the rights and dignity of others.

In examining this conflict, we observe the very meaning of "illness" and "suffering" actively contested by those who assert pride and contentment around their physical condition, challenging as they do the "ableism" of those who would lump all physical conditions that deviate from the norm as inherently defective. Here we also find patient advocates, often parents with children or other loved ones impaired by an accident or affected by an illness, who go on to devote much of their time and energy raising money, lobbying politicians, testifying in front of various audiences, and now actively advancing stem cell research through all of these channels.[10] While there is considerable overlap in the kinds of conditions that affect those on both sides of this struggle, disability activists tend to underemphasize any inherent burden or pain associated with their physical condition, pointing instead to social exclusion, cultural stigma, or lack of access to existing quality-of-life and treatment options as the primary problems to be overcome. By contrast, stem cell advocates tend to overemphasize the inherent biological basis of a condition, highlighting the need for greater economic investment and political support for research to speed the discovery of novel treatments and cures. These competing approaches are routinely distinguished as the *social model* versus *medical model* of disability and disease, respectively.[11]

Stem Cell Battles

One of California's most charismatic stem cell advocates, Don Reed, who was first galvanized into action when his son, Roman Reed, experienced a high school football accident that left him paralyzed, is now arguably the most well-known lay advocate for stem cell research. On the drive to meet with Don Reed in his Fremont, California, home, I passed through glistening suburban neighborhoods and eventually crossed a quaint bridge and drove down an old main street that had an uncanny resemblance to a Wild West Hollywood set.[12] In a sense, the setting mirrored the pioneering ethos of the stem cell advocates who actively campaigned for and now energetically defended Proposition 71—making the drive to meet Reed an exercise in geopolitical metaphor. I could think of no better place for the quintessential pro-cures cowboy to call home.

In many ways, Don Reed epitomizes stem cell advocacy in his weaving together of the drudge work involved in being a lobbyist with the inspirational poetics entailed in representing a sacred cause. He has spent the majority of his waking hours over the last ten years thinking, talking, and writing about finding a cure for spinal cord injury following his son's accident. He has enlisted thousands of people and raised millions of dollars, until finally he was able to get the *Roman Reed Spinal Cord Injury Research Act* passed by the California state legislature, which allocated funds for a research program at the University of California–Irvine, also named after Roman.[13] By his account, these efforts have been absolutely grueling, though his round-the-clock advocacy is not unlike that of other patient advocates who take on this kind of work in the wake of a relative's accident or diagnosis: starting organizations, lobbying government officials, and partnering with researchers in order to advance their quest for a cure. To sustain the labor and sacrifice needed, there is a strong need for visionary discourse and a hopeful imagination, which Reed has mastered. Transporting me from his dining room table to ancient Egypt, Reed embeds his son's experience within a very long history of human injury, masterfully co-opting the "right to life" discourse of his opponents in the process: "The first mention of spinal-cord injury paralysis is on the walls of pictograms of an Egyptian tomb. And it said roughly, of paralyzed soldiers, 'deny them water, let them die, there is nothing that can be done.' And now for the first time, there may be something we can do."[14]

The coupling of the grueling pragmatism that is characteristic of Reed's everyday lobbying with the ardent hopefulness that is characteristic of the "Yes on 71" campaign is evident in the work of a number of stem cell advocates. But Reed's unfaltering commitment to biomedical cures, like that of many of his allies, rests upon a propensity to conceive of any critique of Prop. 71 as originating with "the enemy" and to depict everyone experiencing illness or disability as potential beneficiaries of stem cell cures. As he writes on his blog, "Stem Cell Battles": "We are America's millions: patients, family, and friends. We support research to bring cures, to empty the wheelchairs everywhere."[15] As with Dr. Yamoah's presumption that deaf people are cut off from the social world, Reed's well-meaning vision makes no allowance for what is well documented in social science literature, which is that many "vari-abled" individuals do not view their conditions through the same prism as able-bodied individuals do. Consider the provocative words of disability scholar and activist Gregor Wolbring, who prefers crawling to artificial legs, wheelchairs, *and* walking, from an essay (addressed to those he calls his "non-afflicted" readers) entitled "Confined to Your Legs":

> I did not view my body as deficient and did not see artificial legs as a sensible solution to my primary problem: dealing with a world that saw me first and foremost in terms of my defects, and accorded me so little respect or human dignity that I was not even allowed to choose how I wanted to move around. . . . The medical model

creates the illusion of choice because it internalizes the belief that disabled people are subnormal, and offers science and technology as the solution to that subnormality.[16]

But Reed and other devoted parents—as so many are who battle on the front lines of stem cell research—do not on the surface view loved ones, on whose behalf they labor, disparagingly. Nor does Reed necessarily see the struggle for disability rights as opposed to his quest for cures:

> RB: How do you see that social model, of changing society as op-
> posed to changing our bodies and fixing our bodies, which is
> how [the disability rights perspective] frames it? How do you
> see that competing—
>
> Reed [interjecting]: I feel it is a parallel track. Basically, if I under-
> stand correctly, you're talking about the issue of access. 'Cause
> when we go out to try and get backup for this, sometimes
> people that have been paralyzed for many years, just could not
> bear the thought of a cure existing, and it would just frighten
> them—when you're hanging on by your fingernails, you don't
> welcome earthquakes, even if they're good ones.[17]

In this way, many stem cell advocates support the need for disabled people to have greater access to quality-of-life resources; while at the same time, he and others actively construct and guard the representation of their loved ones' experiences as constituting a species of "personal agony."[18] Despite the fact that Roman Reed appears to live an extremely full life—serving as a planning commissioner for the city of Fremont, for example—it is rare for stem cell advocates to celebrate such accomplishments in their public statements about those on whose behalf they work. In describing to me why some people with disabilities refused to sign the "Yes on 71" petition, for example, Don Reed passionately replied,

> If you cannot move, your life is so changed. Even I—I see my son
> all the time, I see what he goes through—but even I cannot really
> fully understand the agony. Try sometimes to just sit still, and not

move when you want to move. Imagine having to call somebody to move you. Or try this. Take a pen, and try to pick it up using your wrist. That's what my son has to do if he wants to write something. Every single thing is so hard. So you adjust, and over years, you come to learn this very narrow avenue of opportunity that you can do what you have to do, but only just barely. And then have somebody say, "Oh, by the way, there's a possibility that you could be completely, a hundred percent cured." That's insanity to that mind-set.[19]

The empathy that Don Reed expresses and his attempts to persuade listeners rest in part on seeing his son's life as defined by the inability of his body to carry out daily tasks in a "normal" *and* pain-free manner.

The dichotomization of rational versus irrational responses to illness and disability is a related aspect of this quest for cures. For Reed, the experience of being confined to a wheelchair rationally leads one to support SCR.[20] When I challenged him with the idea that many disabled people reject the representation of their lives as defined by perpetual agony and may in fact feel offended by his quest for cures, he referred to this as a kind of false consciousness-cum-"wheelchair rage":

One of my friends went into an independent-living center in Berkeley and said, "We'd like to talk to you about possibly supporting this bill [Prop. 71]," and this guy in his power wheelchair rams her in the shins and says, "You so-and-so, don't you think I'm just as good as you, just as I am?" And the woman [who was soliciting his support] said, "Hey! My son is a quadriplegic." And he said, "No! Get out!" So to me the only answer is, we have to be sure that people who are stuck paralyzed now, and have a chronic disease— which is huge numbers—get all the help they possibly can, that they need now. But at the same time, cure is a part.[21]

Reed's depiction of Prop. 71 opponents as irrational in their rejection and reproach corresponds to feminist disability theorist and activist Liz Crow's assessment that disabled peoples' knowledge is "frequently derided as emotional and therefore lacking validity."[22] It is not coincidental that both Reed and the Prop. 71 petitioner in his story above are well-meaning parents of people "stuck" in wheelchairs—and this kinship bond both motivates and narrows their field of struggle. Disability activist Gregor Wolbring, in turn, reframes the paternalism of patient advocates in these terms:

> The functionality of technologies like the wheelchair is frequently demeaned in expressions like "confined to a wheelchair." . . . No one would use the phrase "confined to natural legs," though in reality people with legs are confined to them, while I can leave my wheelchair when I choose to do so. Nor is the act of driving a car portrayed as "confinement"—instead it is a cultural symbol of empowerment, urban gridlock and stultifying commutes notwithstanding. So the problem is not that technological dependence violates societal norms—modernity is built upon this dependence. The problem is the norms themselves.[23]

Indeed, the image of a man liberated from his wheelchair routinely animates stem cell politics. Don Reed places a wheelchair at the top of his blog, atop a quotation that exults in the promise of stem cell research to "empty wheelchairs everywhere"; similarly, a South Korean postage stamp depicts a person "liberated" from a wheelchair and reunited with a loved one through the power of stem cell research. We see how regenerating the spinal cord can further normalize the image of the *upright citizen* despite decades of legislation and consciousness-raising on the part of disability activists. Thus, developments in the life sciences may not only constitute *new* kinds of political rights but also potentially *erode* the gains that, in this case, disability activists have won.

In truth, the challenge of paralysis is as much social as it is physical, revealing the irreducible tangle between our biological bodies and the many layers of our social environment. Thus, Don Reed routinely reveals intimate details about Roman's everyday experience with spinal cord injury in his public statements, commenting at a stem cell agency board meeting, for example, that "[Roman] had to be picked up and cared for

several times. He had to be cared for—he's not here, so I can say this—
like an infant in many ways. Here's this gigantic noble man, has to go
through hell as part of his daily life."[24] Don Reed's infantilization of his
adult son is part of his rhetorical strategy in the quest for cures, but it
has grave consequences for those whose own quest to be fully respected
"as they are" is made all the more difficult by Reed's constant recapitula-
tion of the "tragedy" of disability. Researchers studying the impact of
disability on family life have challenged the "burdened" family image,
finding "a significant number of parents actually report[ing] numerous
benefits and positive outcomes for their families associated with raising a child with disabilities."[25] This body of work reports that variations
in this experience are not random but typically patterned by income,
age, racial or ethnic status, and gender, among other factors.[26] Working
within an able-ist paradigm, patient advocates like Reed tend to ignore
or downplay this vast and socially structured variation in experiences,
normalizing the tragedy of the so-called fallen outdoorsmen—society
defining people primarily through their actions, their "doing," rather
than their "being."[27] The same sense of tragedy does not tend to characterize those who are *born* with an impairment: it attaches least to those
with a chronic and "invisible" illness (whether congenital or acquired)
and most to those who experience an accident that takes away their ability to move or sense in the same way that they once did.[28]

Interestingly enough, Don Reed's town of Fremont, California, is
also home to a large Deaf community, because the California School
for the Deaf is located there. Historically, Deaf schools have been the
primary site for the cultivation of Deaf community, identity, and empowerment. The Deaf community, in turn, is among the most active in
the disability rights movement, often eschewing the label of "disabled"
altogether in favor of seeking recognition as an oppressed minority, with
its own language and genealogy.[29] Deaf activists, alongside other vari-
abled people who embrace disability identity in order to signal the ways
in which they are not inherently disabled, but rather *made* to appear
disabled by an unjust social system, actively oppose biotechnologies that
offer fetal deselection and regenerative "cures." But just as Fremont's

vibrant Deaf community was largely invisible in my ride through town, in a way that the history of Wild West frontier life was not, so too are disability politics obscured beneath the elaborately constructed frontier image and positioning of stem cell research.

Charitable Scripts and Tragic Crips

To move us beyond the "stem cell battles" as they are conventionally construed and examine the latent struggle between stem cell research and the disability movement, I met with one of the most outspoken disability rights activists in the San Francisco Bay Area, Patricia "Patty" Berne. At the time, she worked at the Center for Genetics and Society (CGS), an organization that promotes the responsible use and regulation of new biotechnologies. CGS's staffers took a very active role against Prop. 71 and, once it passed, attempted to shape its implementation so as to ensure greater oversight of the science and greater accountability with respect to issues of public participation and socioeconomic justice.

Born with muscular dystrophy in the early 1970s to a Haitian father and Japanese mother in San Francisco, Patty Berne described her coming of age in a context of social segregation and multiple forms of discrimination. But as we sat in one of CGS's modest conference rooms on the seventh floor of its downtown Oakland offices, she admitted that she was actually quite fortunate to be brought up in a relatively liberal environment. Not only was her junior high school integrated, "mixed-race and mixed-ability," but at fourteen years old she began working part-time at one of the first independent-living facilities in the country, which she considers to have been a positive experience, "because I generated my own income, which was entertaining for me, and it was amazing for me to work at this place where there were adults with disabilities—that I was not seen in some sort of charity framing, in a rehab framing, in any kind of patronizing way."[30] With the sounds of bustling Oakland traffic whistling through the partially opened office window, along with intermittent rings from Patty's cell phone that interrupt the interview, I am reminded that Berne's full life defies the "tragedy" framing that too often characterizes well-meaning patient advocacy. In our conversation Berne goes on to recall the oppressive underside of altruism as a poster child for the Jerry Lewis Telethon, a fund-raising program for children with muscular dystrophy. Importantly for this discussion of the role of science, medicine, and technology in expanding or limiting human agency, she contrasts her own life desires with the charitable "scripts" that she was expected to follow:

> Berne: The Jerry Lewis Telethon framing was that I would die and live, like, a low quality of life until that point. And [wryly] that wasn't very helpful. Luckily I didn't really hold that as very accurate—there was a lot of lying basically that was involved in the telethon. . . . The charity would get kids with disabilities to go on the telethon and say, like, "Hi, I'm a cute little five-year-old, and I would like to walk, and if you give money, they can do research so I can walk." And they would prompt me [with] lines to say and I would say them—I remember being little, you know, and they were, like, "What do you want to do most in the world?" And I'd say, "I want to ride a motorcycle . . . I want to move fast."
>
> RB: And what did they tell you to say?
>
> Berne: That I want to walk. And I'm, like, No—and I would argue, I would say, "But I don't want to walk; I want to ride a motorcycle." . . . I wasn't down with the messaging they were trying to develop. Still, they don't offer models. A charity framing doesn't offer models for how to live a happy, productive life as a person with a disability.[31]

Berne's rejection of the charitable script mirrors the resistance many in the disability rights movement express toward the promises of stem cell research. Berne's comments also reveal the ways in which her life desires exceed those expected of people assigned to the "handicapped" category, even when well-meaning patient advocates working to "cure" her physical condition must, in the process, ensnare her in that diminished status in order to speak and work on her behalf.[32] Berne and other disability activists' resistance to categories that locate "defects" in their bodies rather than in society illustrate philosopher of science Ian Hacking's assertion that "making up people changes the space of possibilities for personhood. . . . Hence if new modes of description come into being, new possibilities for action come into being."[33] For Berne, the oppressiveness of the "handicapped" label and the low expectations it imposes was learned through everyday encounters:

> When I started going to junior high, I realized the extent to which disability was seen as not just marginalized, but like a condition worse than death. . . . People would say these really off-the-wall things to me. And I really learned, kind of being around non-disabled kids, how the world perceived me—you know, as "less than." . . . There was just a lot of surprise that I had any kind of social graces. There was surprise that I had academic [skills].[34]

The ingrained prejudice Berne experienced from peers as a young person, coupled with her exposure to organizations run by and for people with disabilities, was critical in leading her to become active in the growing Bay Area disability rights movement, which eventually led her to join the staff of the Center for Genetics and Society. Her work at CGS involved engaging with disability rights organizations with respect to the effects of new reproductive and genetics technologies—which had extremely personal implications for Patty Berne, because her own condition, infantile progressive spinal muscular atrophy, is caused by a single gene on chromosome 5; conditions with such clearly defined causes are the target of genetic tests that allow expectant parents to use preimplantation genetic diagnosis to deselect affected embryos or abort them upon learning that a fetus is affected by a particular condition. These practices are precisely what many disability activists consider to be a form of high-tech eugenics, in that people who have a condition that places them outside the species-typical range of physical variability are labeled defective, and their social value is diminished.

As I learned more about Berne's specific job description at CGS and how it involved reaching out to existing disability organizations to raise awareness about the harmful potential of many new biotechnologies, I thought to ask her about how she conceives the difference between the organizations she works with and those patient advocacy organizations that were integral in passing Prop. 71. She observed in reply that

> [a] lot of disease-specific groups, like [the] National Diabetes Association or [whoever], they really see through a medical lens around what the problem in the world is, and people really then advocate for cures and treatment and research. . . . I think that, again, it's where the emphasis is placed. People in disability rights are never saying, like, "Don't do research"—that's absurd. Just like in the same way that people in patient advocacy organizations would never say, "People with disabilities don't have rights." They would never say that.
>
> RB: [So] it's a matter of priority . . . and it's really difficult, if not impossible, to really prioritize both? Or to frame it in some kind of hybrid way?
>
> Berne: Right. I mean, I think it's definitely possible to address healthcare concerns . . . within a disability rights framing—absolutely. It's much harder to address civil rights in a medical framing. How often are civil rights brought up in the medical framing? You know, but certainly civil rights include health access. You know, include people's right to treatment, people's right to basic healthcare.[35]

Interestingly, Berne does not dismiss the medical perspective outright, nor does she completely distance the disability agenda from the quest for cures. She even goes on to admit that she thinks people with disabilities too often adopt a medical perspective, and that her work in many ways requires a purposeful cultivation of an oppositional view of disability *as* social oppression rather than as personal tragedy. Her comments here mirror Don Reed's imputation of false consciousness to wheelchair users who did not sign on to Prop. 71. In both cases, neither holds an essentialist view; rather, they each accept the challenge of educating people about what they consider the best interests of people with disabilities.

Another feature of Berne's discussion of disability consciousness-raising is that she draws a clear parallel between "disability pride" and "black pride," according to which consciousness-raising is central to the effectiveness of *any* civil rights movement. She goes on to explain that she has been reluctant to engage what she terms "patient advocacy organizations" from a disability perspective for two reasons:

> I haven't started working much with patient advocacy organizations, partly because of my own stuff, and partly out of strategy. I mean, strategically it makes sense to organize your allies first; but also, I mean, I was a [Jerry Lewis Telethon] poster child. And I don't feel like being pathologized. And when I read the websites of some of these [disease] organizations . . . I see the way in which I'm not respected. And it's really hard for me to work with people if I don't feel like I'm respected.

> RB: And do you think people with disabilities are at the helm of these [disease advocacy] organizations?

> Berne: There are a lot of people that make money off of disability. There are a lot of institutions that are run for people with disabilities . . . but are run by able-bodied people. And it could be family members; it could be professional advocates. And there are some people with disabilities that still really do use a medical framework.[36]

Interestingly, Berne's reluctance to deride stem cell research advocates is indicative of the absence of overt conflict between the two movements, both of which have a strong following in the San Francisco Bay Area. After all, on the surface, Roman Reed and Patty Berne are somewhat similar: both are young adults making their way within a society that often reduces them to handicapped wheelchair-users. But just the fact that Berne advocates for her own interests, whereas Roman's father, Don, is the main agent on behalf of *Roman's* purported interests, reveals a striking difference with respect to where the locus of agency tends to be in the two movements.

Even so, there exists an uneven playing field tied to the relative political and economic capital that disability and patient advocates have at their disposal. Don and other patient advocates are comfortably situated as the de facto constituency of a people's science, benefiting from the "corporatization of voluntary associations"[37] through a symbiotic partnership with the biotech industry, which in turn benefits from the moral weight of the "fallen outdoorsmen" narrative. This leads patient advocates, as compared with their counterparts in the disability movement, to be much less critical of the biotechnology industry in general, and to place far fewer demands on the stem cell initiative in particular, with respect to the equitable distribution of future biomedical goods. When the subject of intellectual property came up at an Independent Citizens' Oversight Committee meeting, for example, Don Reed, positioning himself as a consumer-in-waiting, as it were, was quick to interject the need to maintain strong incentives for biotech to translate basic research into therapies while limiting the ability of the legislature to demand high royalties:

> The attempt to make this affordable has been tried before. The NIH has tried it, and there's a good study which shows that it was an utter disaster. What people came to realize is that before we can have affordable computers, we must have computers. The greater good is the benefit of this, not the small individual tinkering, which will slow the whole thing down.[38]

Reed's dismissal of attempts to make therapies affordable is, on the surface, less troubling than if we heard the same argument articulated by a venture capitalist who stood to profit from such a laissez-faire approach to stem cell research and development. The spirit of "biocapitalism"[39] he invokes stands in contrast to the economic frameworks typically employed by disability advocates, who tend to foreground affordable access and distributive justice in setting their agenda.

Whose Good Life?

Returning to the framework of bioconstitutionalism, we can observe how judgments about "normal" versus "defective" bodies are layered upon judgments about what constitutes a good versus not so good life, so that the struggle for collective rights to access "the good life" is also sometimes a struggle over what constitutes "the good."[40] From a disability perspective, for example, the proverbial ladder of success is better understood as a hierarchy of *domination* structured according to what we have come to accept as the natural "order of things"[41]—normals on the higher rungs, defects on the lower ones.[42] Regenerating the body, worthwhile as that may at first seem, grows directly out of an able-ist paradigm that defines the good life, and the good citizen who is obligated to pursue that life, in ever narrower, technoscientifically mediated, and often antidemocratic terms. Wolbring makes these connections when he observes that

> [w]hat forms of ableism and favoritism of abilities one exhibits has a direct impact on how one defines and perceives what constitutes a good life, what the problems are that prevent the reaching of that good life and what solutions are thought out to deal with the "problems." The discourses around science and technology governance leave out many facets and subgroups of earth's population. . . . The less diverse the reference group is, the easier it is to define a certain vision of the good life. . . . Social risk, social health issues are rarely raised by the proponent or opponents of a given contested science and technology. And certain groups of earth['s] population are routinely excluded.[43]

As illustrated by the successful passage of the California stem cell initiative, patient advocates like Don Reed, rather than disability activists like Patty Berne, are the de facto public in the eyes of the state, in part because such an initiative seeks to insulate itself from advocacy and lobbying agendas that do not completely support its aims, and disability and other politicized interests pose a threat. What's more, prior to the "Yes on 71" campaign, California-based disease advocates who eventually became central to the stem cell initiative already had firsthand experience crafting relationships with state officials, biomedical research centers, and private firms. While Don Reed's level of engagement and effectiveness is certainly exceptional, scholars have pointed to a number of remarkable collaborations between lay advocates, who are animated by personal experiences, and their powerful institutional allies.[44] Among the many national organizations that either have been formed in response to advances in stem cell research or have shifted their disease advocacy energies to this arena, the Michael J. Fox Foundation for Parkinson's Research, the Juvenile Diabetes Research Foundation, the Parkinson's Action Network, and the Stem Cell Action Network were particularly instrumental in the passage of Prop. 71. All of these are in addition to the powerful institutional centers that conduct research and which partner with patient advocates in fund-raising efforts and to secure positive media exposure, insofar as advocates reiterate and reinforce the primary moral imperative underlying this otherwise controversial field: to end physical suffering.

What's more, with ten of the twenty-nine CIRM board seats set aside for representatives of particular disease conditions, patient advocates have become essentially the de jure public with respect to stem cell research as well.[45] Providing a cogent articulation of the central understanding that underlies the work of stem cell advocacy, Bernard Siegel, the founder and executive director of a pro-cures organization called the Genetics Policy Institute, energized attendees at the 2008 "State of Stem Cell Advocacy" conference by affirming that "[s]tem cell advocacy is not a political movement. It is a consumer movement. If you ask people on the street if they support this, they do—not because it is a public health issue—but because it's a personal health issue."[46]

Stem cell advocates are concerned with expanding and protecting a consumer-based liberalism, ensuring access to future biomedical goods and services, and in that way they are very similar to other public interest and citizen advocacy groups that have been ascendant for some time. In one study of this trend, scholars describe a "postmaterialist" liberalism thriving in the civic sphere, increasingly focused on issues that appeal to their middle-class supporters, which "have become less likely over time to ally with traditional liberal groups on behalf of redistributive social programs."[47] This "upwardly tilted public agenda"[48] echoes throughout the stem cell frontier. Siegel's remarks resonate with the conference motto, "Families for Cures," which astutely challenges the pro-life movement's opposition to SCR on its home turf of "family values."

Against the backdrop of the controversial national debate around embryo politics, Robert Klein, the initiative's main architect, saw to the crafting of board membership and the dynamics of working-group participation so as to closely guard the extent to which critical debate might characterize the Independent Citizens' Oversight Committee's deliberations.[49] The claims of stem cell advocates that their efforts are "apolitical" run counter, of course, to the practice of formulating stem cell governance on the basis of potential board members' loyalty to the cause of stem cell research. On the contrary: the stifling of robust political debate rests on a political judgment regarding how far the perimeter of "citizen oversight" should be allowed to extend. It is for this reason that, contrary to the usual concerns of policy watchdog groups who focus on the stock portfolios of board members, I contend that a *lack* of conflicts of interest—namely, the exclusion of disability perspectives— poses the more serious threat to democratic participation with regard to stem cell research.

That said, it is important to point out that there is no *one* "disability perspective" which could be brought to bear on the California initiative by simply bringing the right combination of spokespeople together (as was attempted with a "CIRM Diversity Focus Group" that mainly sought to increase support for the agency among racial and ethnic minority populations). Irvine, California–based Deaf activist Richard Roehm, for example, welcomes stem cell research, although he acknowledges that "the majority of the adult deaf would perceive this as some sort of genocide."[50] Others with disabilities, like rapper Richard Gaskin (Professir X), who was paralyzed after a gunshot wound to his spine, have become spokesmen for stem cell research, performing odes to Christopher Reeve and singing stem cell anthems that ask for "the wheelchair to be a thing of the past." Still others, like blogger John D. Walker, complicate Deaf pride politics by questioning the meaning of biotechnology for Deaf identity through the prism of hybridity. He argues that technological enhancement is not a threat but rather a welcome instrument to enhance human existence, and suggests that Deaf identity shouldn't rest on opposing "hearing identity": "[W]e exist because we are. As long as the affinity exists between us, the Deaf community exists."[51] But despite the lack of any uniform perspective with respect to stem cell research on the part of the disability activists, it is vital that the full range of issues and concerns that people with disabilities wrestle with, as opposed to only the ones narrowly defined in terms of a medical model, gain a genuine hearing in the implementation of initiatives that will directly affect and implicate them.

Consumer Rights Without Civil Rights

Both stem cell advocates and disability activists share a focus on bodies and rights as the two-sided loci of their work, framing their concerns and interventions in terms of the medical versus social models of disease, respectively. For many stem cell advocates, such as Don Reed, the challenge is to remove political obstacles like the federal ban on using human embryos in research, so that people who experience illness and impairments can access the right to transform their bodies in very fundamental ways—such as, to pick a dramatic example, walking if they are paralyzed. On the other hand, for many disability activists, such as Patty Berne, the challenge is to reconstitute the socioeconomic landscape so that people who experience illness and impairments can obtain access to workplace and housing accommodations, along with greater recognition and respect for what they have to offer society. For both communities, the struggle over these different rights is directly tied to how they regard the source of their bodily suffering—that is, as either biologically inherent or imposed by society—and where they draw the line between human diversity and human defect.

Yet, the patient advocates at the helm of California state politics rarely have to seriously contend with the perspectives of people with disabilities. The prospect of Deaf Crusaders bombing a stem cell lab or taking the stem cell agency to court for violating their civil rights as an oppressed minority is not on the horizon of concerns for the pro-cures movement. For Reed and his allies, the "stem cell battle" is largely a face-off against religious conservatives whose pro-life political agenda undermines the right of the living to seek stem cell cures. The conflict between the "Yes on 71" campaign petitioner and the independent-living facility resident in Reed's account lays bare the usually latent conflict between disability activists and stem cell advocates. Those like Reed and other Prop. 71 proponents who have spent the last decade engaged in a struggle to have their quest for cures prioritized by governments, supported by the general public, and funded by commercial entities and philanthropic foundations now have an "insider status in scientific controversies."[52] In the words of the California stem cell initiative's main architect, Robert Klein,

> In California, people were quite receptive to understanding [stem cell research] in the context that this is a solid area in family values where families have historically been given the right to access and to have the country develop the best medical technologies and therapies for their children, their aging parents, and their spouses. And to prohibit the development of these therapies is to really prohibit parents from having the choice of accessing therapies for their children [and] their aging parents. It's really a government intervention in the rights of the family to get the best healthcare in this nation. And it's taking away rights that have been sacrosanct and held by the family for a very long time in this country.[53]

The work of Siegel, Klein, and Reed, among many others, represents a new turn in patient advocacy, galvanized in part by the promise of SCR. In contrast to the loosely networked associations that have typically characterized the activities of patient advocacy organizations as they pursue their distinct agendas, the newfound necessity to defend the "right" to conduct SCR vis-à-vis restrictive federal policies has coalesced patient advocates' activities around a quest for biomedical cures.[54] It is precisely within this umbrella "pro-cures" agenda that patient advocates make claims on behalf of "all those suffering" with an illness or impairment, which makes their assertions so powerful in the court of public opinion.

In contrast to this consumer rights agenda, those in the disability rights movement have spent several decades struggling for basic *civil* rights—access to employment, education, and housing—*and* for social inclusion, against the backdrop of forced institutionalization and eugenic schemes, more tacit prejudice and philanthropic campaigns, and most recently, "high-tech" projects such as Prop. 71. It is in this context that Patty Berne expressed that

> [w]hen there are millions of people in the U.S. that don't have access to healthcare, when the majority of people with disabilities live below the poverty line, to put so much resources into cures that no one with a disability actually—well, all the ones I know of—very few people with disabilities could actually afford. . . . I think that if people want to be visionary about healthcare, there needs to be universal healthcare. . . . I mean, but quite honestly there's a huge leap of logic, and I would be hard-pressed to think of working-class people that have that vision.[55]

Disability advocates' demands tend to range from pragmatic concerns like Patty Berne's about universal healthcare, to democratic concerns about ensuring that people with disabilities are fully integrated within the governing structures of biotech initiatives, to more ontological concerns about what physical conditions are labeled defects as opposed to being considered part of the spectrum of human diversity. Within this range of concerns, there is an underlying focus on the lack of inclusion of disability perspectives within the development of science and technology more broadly—hence the popular movement motto "Nothing about us, without us."

The unforeseen effects of the newly coalesced pro-cures agenda are quietly becoming a concern for some in the biomedical and bioethical fields. As the ties between commercial entities and volunteer organizations grow stronger, patient advocates' conflicts of interest[56] and lobbying power[57] have come under greater scrutiny.[58] For example, Mildred K. Cho, associate director of the Stanford Center for Biomedical Ethics, "wonders about the long-term effect on the culture of science when research is funded by popular vote. She points to the patient-driven factions arising around autism research, in which parent groups attack study results they don't like and raise money to test their own treatment theories. 'The public drives not just what disease areas get attention but what the research strategies are,' she warns."[59]

But not only has the factional character of patient advocacy been effectively subsumed under a broader agenda; patient advocates are now typically understood to be allies and instruments in furthering biotech developments. Thus the counterpart to Cho's concerns relates to the long-term effect on processes of political engagement when a powerful composite of disease organizations essentially pushes the disability perspective and its social justice agenda "off the map" of the stem cell frontier.[60] As articulated by one advocate: "Disability strategies are not permitted to address issues of bioethics. . . . 'the catastrophe of disability' and the hope for a cure remains not only the story of disability in our culture; it is in a very deep sense a governing myth and grand narrative. As such, it is a myth with a profound influence on the possibilities of democratic participation for people with disabilities."[61] That narrative is not only a myth that undermines the dignity and personhood of many disabled people, the majority of whom do not accept the inherently pitiable images that dominate mainstream representations of disability. The consumer-based liberalism of patient advocates also neglects the reality of many people with illness and disabilities who are cut off from the most basic public goods—housing, employment, healthcare—as a result of their subordinate socioeconomic status. As with other bioconstitutional struggles, depending on *who* one imagines "the people" of this initiative to be—individuals with biological defects who primarily desire a cure, or a vari-abled populous that seeks full inclusion—*how* one seeks to implement a "people's science" shifts markedly. In one vivid account, Marci Roth of the Spinal Cord Injury Association testified before Congress following the Hurricane Katrina disaster:

> [On August 29] Susan Daniels called me to enlist my help because her sister in-law, a quadriplegic woman in New Orleans, had been unsuccessfully trying to evacuate to the Superdome for two days. . . . It was clear that this woman, Benilda Caixetta, was not being

evacuated. I stayed on the phone with Benilda, for the most part of the day. . . . She kept telling me she'd been calling for a ride to the Superdome since Saturday; but, despite promises, no one came. *The very same paratransit system that people can't rely on in good weather is what was being relied on in the evacuation.* . . . I was on the phone with Benilda when she told me, with panic in her voice, "the water is rushing in." And then her phone went dead. We learned five days later that she had been found in her apartment dead, floating next to her wheelchair. . . . *Benilda did not have to drown.*[62]

Benilda Caixetta's story is not one of an inherent tragedy resulting simply from a biological defect. Rather, it is a much larger story about the tragedy of socioeconomic inequity, preceding and then exacerbated by a hurricane. It is a tragedy that would not have yielded a different outcome in a world of regenerated cells—because, after all, those living on higher ground will always have access to expensive treatments, while the majority of citizens will not. The real genius of Prop. 71's populist campaign was to obscure the social landscape, so that California voters came to believe that regenerative medicine would airlift them out of inevitable morbidity and mortality.

CHAPTER THREE

EGGS FOR SALE

I'm pro-choice, but when did the right for women to control their own bodies turn into a laboratory's right to experiment with human embryos?
—Diane Beeson, Pro-Choice Alliance for Responsible Research

Why should everyone else in the stem cell assembly line be compensated for their contributions to research, while women are expected to be altruistic and donate without compensation?
—Legal scholar Radhika Rao[1]

SINCE 2009, Jennifer Lahl, the writer and director of the award-winning documentary *Eggsploitation*, has engaged in a national campaign to raise awareness of the exploitation of young women by the fertility industry—a campaign that has direct implications for stem cell researchers seeking "leftover" embryos. Upon completion of fertility treatment, in vitro fertilization (IVF) patients can typically choose whether to donate the leftover embryos to another couple, have them discarded, store them indefinitely, or *donate them to research*. This IVF excess occurs when more eggs are derived through ovarian hyperstimulation than can safely be implanted in a woman's uterus without the potential for multiple pregnancies. Many observers worry that as pressure and incentives mount for doctors to procure more eggs to "share" with researchers, the risks for women increase as well. Until very recently, with the establishment of the Infertility Family Research Registry (IFRR), there was not even any systematic mechanism to collect long-term data to inform women who are considering provision of their eggs for either research or IVF purposes.[2]

Since the release of *Eggsploitation*, Lahl has continued to report on the experiences of women, who are coming forward with their experiences of egg harvesting. Among these is Shavonne, who was twenty-eight years old at the time she responded to an ad in her local paper for "African-American egg donors" which offered $6,000 for eggs. Motivated by the money, Shavonne went to the clinic, where she was the only prospective donor at the time who asked a lot of questions, she reports, which seemed to annoy the clinic staff. Even so, she decided to undergo the procedure. Soon after, however, the receiving couple "changed their mind and no longer wanted [her] eggs." At this point the clinic staff asked Shavonne if she would "be willing to donate [her] eggs to embryonic stem cell research," which she agreed to do "because I didn't mind them being used for that."[3] So on Thanksgiving Day, 2006, Shavonne took a drug called Follistim that stimulates ovulation, which is when her experience took a turn for the worse:

> The retrieval went fine, but not too long after that my stomach started to swell, and every time I leaned over I could feel my ovaries "plop." I went to see the doctor, and he told me I had OHSS [Ovarian Hyperstimulation Syndrome], and he then said, "We see girls like you all the time." I looked 4 months pregnant. They told me to go home and eat a lot of protein. My mother was staying with me at the time, and one night my stomach was so swollen and I could hardly breathe. My mother said, "That's enough," and took me to the emergency room. The nurse stuck a needle in my stomach, and it was a loud pop I could feel, like a balloon was popped. She stuck a bag on the end of the needle to drain the fluid, and the bag filled with 2 quarts in about 5 minutes. She had to quickly put another bag on and some of the fluid spilled on the floor. She filled the next bag too—in all, 4 quarts were drained out of my stomach. I stayed in the hospital for 2½ more days while they drained more fluid. I had a lot of pain in my abdomen. The staff at the hospital would shake their head at me and took pity on me, because I was an egg donor and they said they saw this a lot.[4]

In addition to the immediate physical toll that the ovarian hyperstimu-lation exacted on Shavonne, she was burdened with huge medical bills that took a year and a half to pay. Her own fertility has been compro-mised owing to irregular menstrual cycles; she became pregnant in 2008 and miscarried.

Experiences like the one Shavonne reports energize feminist activ-ism against the commodification of women's eggs. Medical sociolo-gist and member of the Pro-Choice Alliance for Responsible Research, Diane Beeson, and others argue that were it not for the financial in-centive, Shavonne would have been able to consider the risks involved in egg harvesting more soberly. Such concerns motivate the dominant feminist argument against compensation in both public and private research contexts in California, even if such bans are unlikely to be adopted in the fertility industry.

In recent years, public awareness about the role of women in the assisted reproductive technology (ART) industry has increased with the proliferation of newspaper articles, documentary films, and books that explore the social and economic factors that structure this arena.[5] As a set of exchanges that occur in the private sector and draw in part upon a select demographic of (young, mostly white, and educated) egg suppliers, much of this buzz tends to emphasize the prerogatives of the parties involved. By contrast, normative prescriptions and regu-latory proposals abound when it comes to the tissue economy upon which stem cell research depends, while at times little attention is paid to the interests and agency of women.[6] The pool of egg suppli-ers for stem cell research is also more socially diverse in terms of race, age, class, and education than for IVF suppliers because prospective parents tend to be much more selective with respect to these demo-graphic characteristics than researchers. Greater heterogeneity in the research context animates the concern both of those who advocate greater top-down regulation and of those who seek a more horizon-tal contract relationship between researchers and egg suppliers in the stem cell arena.

Eggonomics

In 2005, a handful of researchers around the world, including those at the University of California, San Francisco, Stemagen (a private company in La Jolla, California), and the Harvard Stem Cell Institute, were excited about a promising new method to produce genetically tailored embryonic stem cells. Somatic cell nuclear transfer (hereafter SCNT) is a technique in which the nucleus from an egg cell is replaced with the nucleus of a donor's somatic (nonreproductive or stem) cell so that when it's stimulated to divide, the new hybrid contains the genetic information of the donor. Also called "therapeutic cloning," SCNT allows researchers to develop tailored stem cell treatments that won't be rejected by a recipient's immune system.[7]

In addition to the myriad technical difficulties involved in honing SCNT, researchers also ran into a serious "nontechnical" challenge: to hone the method, they needed a large supply of eggs, because it takes multiple attempts to create viable embryos using this method. While the initial excitement about SCNT has waned somewhat in the wake of newer techniques to produce pluripotent stem cells (cells which can differentiate into different cell types), researchers still insist that a steady supply of eggs is necessary, because they have yet to confirm whether induced pluripotent stem cells (iPSCs) contain all the properties found in embryonic stem cells.[8] But in the larger context of egg supply and demand, how can researchers compete with the private in vitro fertilization arena, which attracts egg suppliers with substantial monetary compensation?

Harvard University's Kevin Eggan is among those who utilize SCNT in trying to learn more about the degenerative neurological disorder called ALS (or Lou Gehrig's disease) in the hopes of eventually producing a cure. He and other stem cell researchers initially thought that spare embryos from fertility clinics could serve as a ready supply. Indeed, the University of Wisconsin lab which first isolated embryonic stem cells obtained their supply of embryos from neighboring clinics. According to estimates from 2003, over 400,000 embryos were being stored in the United States,[9] with hundreds of thousands more in other countries,[10] providing an initial storehouse for stem cell research. So the initial expectation of many in the field was that excess from the private fertility industry could be recycled in the public research context, serving as a ready source of tissue.

But by 2008 many researchers, including Eggan, began publicly stating that this source was not adequate, on account of deficiencies with respect to both quantity and quality of the embryos stored in IVF freezers. At a meeting of the California stem cell agency, Eggan testified that he and his colleagues had spent over $100,000 advertising for egg suppliers because the IVF supply was insufficient; however, he reported, they had little success in recruiting women to provide eggs "for free" when they could donate in the IVF context and make upwards of $5,000.[11] He also made his case in newspaper articles with headlines such as "Stem Cell Researchers Warn of Egg Shortage in the United States" and "An Egg Shortage: Is More Cash the Answer?" His public testimonies were an effort to get the California agency, federal research agencies, and other public stem cell initiatives that are modeled after California's to reconsider their strict noncompensation policies for egg suppliers.[12] From the perspective of Eggan and his colleagues, noncompensation policies in the public sphere, competing alongside the laissez-faire exchange of tissue for payment in the private IVF sector, are impeding the forward march of regenerative medicine. The inability to redirect IVF egg suppliers—notably young, white, educated women whose demographic characteristics make them highly sought after by infertility patients—from the private commercial sector to the public research sector is a growing issue for the stem cell field.[13]

Many proponents of stem cell research initially took for granted that they would need to develop incentives for redirecting egg suppliers from private reproduction to public experimentation. But with financial compensation off the table in most jurisdictions, other, nonfinancial techniques become necessary. In this context, it is no longer enough to ensure ethical integrity and seeing to the medical safety of individuals. Rather, those implementing public initiatives must increasingly grapple with the socioeconomic status and political rights of tissue suppliers as a precondition to their participation. These considerations are not contaminants of pure science but a necessary part of establishing the niche in which public science can potentially thrive.

Framing Feminisms

While stem cell research is characterized by the same tissue procurement methods and risks encountered in ART clinics, the framework of "people's science" leads to a very different conception of women's rights, a notion that is deeply contested, when it is not ignored. This is exemplified in a September 2006 *Los Angeles Times* headline "New Battle Lines Are Drawn Over Egg Donation," saying that "feminists are split" over whether to pay women.[14] The organization Pro-Choice Alliance for Responsible Research was among those who worked tirelessly in the months leading up to and following Proposition 71 to oppose the California Stem Cell Research and Cures Initiative, largely focusing on the uncertain health risks that multiple-egg extraction poses for women. Medical sociologist Diane Beeson testified before the United States Congress on May 7, 2006, warning that

> [a]s a society, we are at a turning point. . . . We are being asked
> to make women the servants of biotechnology rather than insisting
> on a biotechnology that promotes the well-being of all people. . . .
> Until we understand more fully its human costs, I strongly urge
> your support for a moratorium on [somatic cell nuclear transfer] in
> both [publicly and privately funded contexts]. (Available at: http://
> handsoffourovaries.com/images/beesontestimony.pdf.)

Bringing together people across the political spectrum in an unprecedented way, the Alliance was supported by the California Nurses Association, based in Oakland, California; the National Women's Health Network in Washington, DC; and Our Bodies Ourselves, the Boston-based women's health advocacy group. Beeson and her anticompensation allies joined together in a campaign called Hands Off Our Ovaries that warned of the possible dangers of egg donation for research purposes. For example, problems associated with Lupron, the drug most often used to shut down a woman's ovaries before stimulating them, include severe joint pain, difficulty breathing, chest pain, nausea, depression, emotional instability, loss of libido, severe weakness due to loss of pituitary function, amnesia, and hypertension, among many others.[15] In severe cases of Ovarian Hyperstimulation Syndrome, women experience cysts, fluid buildup, and enlargement of the ovaries, which for many of those concerned are all the more troubling because egg extraction is a "non-vital treatment with a potentially fatal outcome."[16] Another aspect of the risks associated with egg extraction is the uncertainty about the long-term effects of multiple-egg extraction; egg suppliers report health problems such as memory loss, bone aches, seizures, and vision problems that began when they were taking Lupron and have not gone away.[17] Yet technically, the clinics (or stem cell labs) that keep such close watch over a woman's health for the duration of the procedure are no longer responsible for their long-term care.[18]

Many feminists therefore call for a moratorium on the use of eggs in research "until we understand its human costs more fully."[19] Of major concern is the laissez-faire approach to paying women for eggs in the fertility industry and how that may transfer over to the research context, for example coercing many women to undergo invasive procedures with known short-term risks and many unknown long-term risks. Rather than offset the risks with some form of compensation, as occurs (in the thousands of dollars) in the fertility industry, Beeson and allies have worked to uphold a strict "noncompensation" policy in the research context as a way to dissuade women from undergoing the risky procedure. This policy was in fact codified in Prop. 71; the text of the initiative seeks to ensure that egg donation is in no way coercive, thereby instituting a "prohibition against compensation," albeit one that allows reimbursement for all donation-related expenses.[20]

Still, stem cell researchers have found loopholes around California's noncompensation policy; and there is also little consensus among feminists about whether a noncompensation policy is actually in the best interests of women. Some, like the Feminist Majority and the California chapter of the National Organization of Women (NOW), lent their support to Prop. 71 on the basis of a general optimism about the health promises of stem cell research. "How can we turn our back," argued Helen Grieco, executive director of California NOW, "on the potential of this groundbreaking scientific research in good conscience? As a feminist organization, we support health initiatives that benefit women. Stem cell research benefits all of us."[21] Grieco's optimism resonates with the efforts of some feminists to position women as beneficiaries of and partners in mainstream scientific and medical initiatives.

The prohibition on compensation at the U.S. federal and state levels, as well as in other nations such as the United Kingdom, Canada, and Japan, is explained in the National Academy of Sciences guidelines in the following terms: "The explanation of such unanimity might lie in the view that the treatment of the developing human embryo as an entity deserving of respect may be undermined by the introduction of a commercial motive into the solicitation or donation of fetal or embryonic tissue for research purposes."[22] We should keep in mind, however, that this prohibition against paying for embryos does not apply in the private sector, where compensation ranging from $5,000 to $30,000 for egg suppliers is standard practice. Presumably the difference reflects a shift from clinical to research uses of reproductive tissue, accompanied by a shift in regulatory framework as well. But what about privately funded research?

In part as a result of feminist activism, the California Senate also passed the Reproductive Health and Research Bill,[23] which expands the noncompensation policy to the private sphere when biotech companies are engaged in stem cell research.[24] Marcy Darnovsky of the Center for Genetics and Society sees the importance of the bill as "defin[ing] women who provide eggs for research as 'research subjects,' triggering federal and state regulatory protections. . . . In order to head off the emergence of a market in which predominantly poor women are the ones who wind up selling their eggs, it limits payment to reimbursement for direct expenses."[25] The ban against commercializing human tissue draws on the notion of donating part of one's body as a "gift" which harkens back to mass blood donation drives during World War II and which enshrines the Western cultural repulsion toward bodily commodification.[26] Blood drives, now commonplace in many societies, serve to ritualize citizenship, communalize health, and animate Anderson's (1991) notion of an "imagined community" in which citizens are literally bonded by blood.[27] The emergence of tissue donation has thus served to unhinge the "natural" distinction between (pure) tissue donation and (profane) tissue commodification—upon which noncompensation activism relies.[28]

Other scholars and activists doubly critique the dominant feminist noncompensation agenda: they remain skeptical about the promise of stem cell research *and* challenge the conflation of egg donor compensation with exploitation; they advocate greater reciprocity between researchers and egg donors in the form of financial and nonfinancial forms of recompense for women's biolabor. They also seek to loosen the link between social justice and greater public regulation, allowing for the possibility that women's interests may be best served if bodily commodification is made mutually beneficial to researchers and donors within the growing "tissue economy." This is the approach of San Francisco-based legal scholar Radhika Rao, who asks, "Why should everyone else in the stem cell assembly line (researchers, companies, the state, etc.) be compensated for their contributions to research, while women are expected to be altruistic and donate without compensation?"[29] She and others argue that women are deserving of some form of compensation insofar as their biolabor is fundamental to the research process. Similarly, UC Berkeley sociologist Charis Thompson compares egg donation to "other kinds of physically demanding service work," arguing for a "salary negotiation between the state agency (or relevant employer) and the donor." This, Thompson contends, is a "sensible and dignified recognition of [the donor's] work, time, and effort." Responding to the question of whether donor compensation is necessarily coercive, Thompson notes that there is no reason to think that *emotional* inducements, like having a loved one who can potentially be treated by stem cell therapies, "are any less coercive or less likely to cloud motivation or the context of informed consent."[30]

This feminist school of thought, which focuses on ways to enact reciprocation in the research process, suggests that the extent to which bodily commodification is exploitative is context-dependent rather than a universal political calculus that equates compensation with coerciveness. Legal scholar Michele Goodwin captures this approach well when she says we need to "change the discussion from whether or not to commodify, to what degree of commodification is socially acceptable."[31] Criticizing the laissez-faire regulatory approach to egg donation in the fertility industry, those advocating a reciprocal exchange offer a more discriminate and pragmatic approach to ensuring women's safety rather than a wholesale moratorium against egg research. Donna Dickenson puts it this way: "If the lady is not to vanish altogether, she needs protections such as contract, but a form of contract limited to the protections she most needs, and aimed at ensuring that women's contribution to stem cell technologies is actually recognized. . . . Without [such] recognition . . . the terms in which the stem cell debate is conducted are themselves deceptive and disingenuous."[32] For Dickenson and others, institutionalizing a reciprocal relationship between researchers and egg donors via a contract focuses on the protections women most need rather than eliminating the option to donate eggs for research outright.

The distinction I draw does not mean to suggest that those primarily advocating protection against harm are not then interested in women's agency, or that those seeking a reciprocal exchange between women and researchers are not concerned with potential harm done to women, or even that people voicing arguments that can be situated in either camp are necessarily animated primarily by feminist concerns. Rather, my primary aim is to illuminate this tension through the prism of bioconstitutionalism: we see that depending on *who* one imagines women are in this context—bioworkers who supply eggs for research, or a protected class of female research subjects—*how* one seeks to implement ethical research shifts dramatically. As with other bioconstitutional struggles over what rights people are owed vis-à-vis life sciences, this one is bound up with the question of who "the people" are thought to be in the first place.[33]

The impulse to protect potential egg donors fueled the Proposition 71 egg donor guidelines, the Hands Off Our Ovaries campaign, and the Reproductive Health and Research Bill (SB 1260). The effort to regulate egg donation in the research context by foreclosing the possibility of compensation was based on the idea that the donors were primarily young, working-class women of color; it stands uneasily against the hefty compensation allowed in the fertility context, where the majority of donors are young, educated white women. The politics of payment, in other words, is complicated by the politics of difference, because different demographics of women are assumed to supply tissue for stem cell research versus IVF. Both camps routinely refer to existing stratification—by economic status, race and ethnicity, age, and disability or illness status—to argue that particular categories of women are uniquely vulnerable to, or uniquely deserving of inclusion in, research. In this way, the social value of particular categories of women and the relative value of their eggs become entangled, complicating the effort of both feminist camps to "speak for women."

For some, the inclusion of racial and ethnic minority women as egg donors is considered necessary to ensure a genetically heterogeneous sample of stem cell lines, so that future therapies derived from these lines will be accessible to minority patients.[34] The inclusion of economically poor women, by contrast, is considered risky, because they are likely to be in poor health and prone to complications associated with egg donation. Much of the tension in the discussion stems from the racialization of poverty that is endemic to the United States, such that conflicting policies cohere around a doubly subordinate subgroup of women. Thus we find that a focus on racial/ethnic and class differences among possible research participants emerges as a central focus in the California stem cell agency's deliberations.

Efforts to protect women from the risks of egg extraction are animated by historical precedents in which poor women (and, owing to their disproportionate poverty, women of color) have been vulnerable to coercion by researchers. The health risks that may result from egg donation led Susan Fogel of the Pro-Choice Alliance for Responsible

Research to ask rhetorically, "How much money is enough to coerce a poor woman? And do we up the ante until they bite?" The relative social worth of different demographics of women comes up against a tissue economy that, advocates warn, is intrinsically indifferent to the welfare of egg suppliers. Thus, as that tissue economy seeks to cut costs, those women willing to accept the least compensation will be induced to submit to the invasive process of egg extraction. Reflecting upon the demand for human eggs in stem cell research, sociologist Barbara Katz Rothman observes that "whether we are looking at the marketing of tissues or of services, it is hardly surprising that price reflects class/race/status. The kind of social revolution we need to make all women of equal 'worth' goes way beyond this particular form of [egg donor] exploitation."[35] Rothman believes that women's interests and the interests of research are intrinsically at odds, and that the social devaluation of particular kinds of women will necessarily transfer into the research context. Her analysis is used by those opposed to compensation to argue that payment is necessarily coercive, particularly for women of limited means and choices.

Those seeking a moratorium on egg research articulate their position as part of a broader critique of reproductive politics, situating themselves on the side of socially subordinate women whose concerns (stemming back to eugenics and state-sponsored antinatalism) typically remain marginal to the pro-choice movement. The ideal of reciprocation between researchers and donors, as well as a broader optimism with regard to the possibilities of stem cell treatment, may gloss over the ways in which doctors and scientists have coerced and abused already marginalized women throughout much of U.S. history. That one of the main failures of reproductive politics to date is the way it upholds the interests of white, wealthy women fuels protective concerns, such as that expressed by outspoken noncompensation proponent Marcy Darnovsky:

> The conditions that shape women's ability to make choices fall out
> of the picture for some, not all, reproductive rights organizations.
> And then the social implications of individual decisions . . . there's

really nowhere that they can get considered and even no vocabulary to talk about it. . . . One of the groups that has felt this lack, this kind of skewing, most strongly have been women of color, and women of color organizations. The right to be a mother, for example, sometimes falls out of the conversations.[36]

The desire to take account of the interests of women of color potentially conflicts with the desire to include the eggs of diverse kinds of women in stem cell research. For example, Goodwin questions the supposed harm that bodily commodification presents to racial and ethnic minorities, arguing that protective policies limit the supply of biodiverse tissue by relying so heavily on the norm of altruistic giving. While her work focuses on organ donation, the same logic is applicable to stem cell research, where some researchers think that if we do not have a diverse egg donor pool, then future stem cell therapies may not be available to many elements of a racially and ethnically diverse population.

Those seeking to establish a reciprocal exchange between egg suppliers and researchers argue that seeking a moratorium on egg donation and/or opposing compensation for donors are not the only or best ways to counter the exploitative possibilities of egg donation. Thompson urges that "instead of refusing compensation to women who donate their eggs so as to ensure that their assessment of risk is sufficiently unclouded, why not direct our efforts at understanding and minimizing the risks so that we dramatically reduce what has to be offset?"[37] Withholding compensation, Thompson argues, is a bad strategy for ensuring safety if that is actually the aim. As she explains, a purely altruistic model can put women at *more* risk: "If the donor pool is restricted by a lack of financial incentive, pressure to donate will be increased on those who stand to benefit in other ways," including those who are affected by an illness that might benefit from stem cell research.[38]

Some, like Rao, posit that protective worries directed at poor women of color are in themselves problematic: "Shouldn't we be worried about the women donating eggs to fertility clinics? If you pay women a lot and they're white, it isn't exploitation?"[39] Similarly, Goodwin is wary about the use of African Americans as "a transitional good in these intense debates," noting that the "most poignant critiques against organ selling are race-based, particularly the claim that commodification would harm racial minorities."[40] Her work attempts to "uncloak the notion that public regulation always benefits the disenfranchised," arguing instead for "a system that allows incentives to coincide with altruistic donation."[41] Goodwin's position resonates with that of Thompson, who draws from data taken in the IVF context to suggest that financial and altruistic motives, "far from being incompatible, seem to bolster one another. . . . It is wrong then to worry that being paid substitutes a financial for an altruistic motivation."[42] The protective inclination to shield poor women, and women of color, from the bad consequences of their limited choices is viewed by some seeking reciprocation as not simply bad ideology but bad ethics as well.[43]

In sum, for both feminist camps invocation of categories of women who stand to lose or gain in the tissue economy reveals the pliability of race and class in disagreements over how to produce eggs for research and justly distribute their potential value. In part, this focus is unique to the state-mandated bases of the California initiative, which make such accountability necessary in "people's science"; but it is also more broadly symptomatic of the racialized class politics of the United States. To the extent that controversial science in other state and national jurisdictions "goes public,"[44] California provides a window into the bioconstitutional struggle to protect and empower particular groups of people—both biologically and politically defined—who will emerge according to the specific axes of stratification found in those other contexts.

In the broader effort to set up statewide regulations for publicly funded stem cell research, three interrelated debates around egg provenance and procurement arose in the California stem cell agency's deliberations in the course of three meetings in 2006 and 2007. The first of these debates was the underlying tension between participation in research as a "risk and burden" to donors versus participation as a "right and benefit" that must be extended to all (female) citizens. The second debate related to competing conceptions of participation as either "work," for which people are motivated by financial incentive, or "service," for which people are motivated by feelings of altruism. The third and final debate concerned the source of women's differences, as either genetically determined or socioeconomically induced, and the implications in either case for access to, and the efficacy of, stem cell therapies. Often explicit in all three struggles was an attempt to address the question of "what kinds" of women should supply eggs for stem cell research.

Risk Versus Right

We can see how the competing logics of the feminist debate described above infused the deliberations of CIRM's Ethical Standards Committee as it debated the issue of reimbursing egg suppliers for direct expenses associated with their donation (as opposed to compensating them more generally). The discussants at a working group meeting in January 2006 included Bernard Lo, professor of medicine and director of the Program in Medical Ethics at the University of California, San Francisco, who is also a member of the Institute of Medicine and the National Academy of Sciences, and who chairs the Standards Committee; Patricia King, professor of law, medicine, and public policy at Georgetown University, who is a member too of the Institute of Medicine and has served on numerous federal bioethics commissions, including those related to recombinant DNA and human genome research; and Sherry Lansing, a well-known philanthropist and the director of the Sherry Lansing Foundation, which primarily supports cancer research. As a University of California regent and a former movie producer, Lansing is also well connected among California's political and Hollywood elites.

Leading up to the meeting, committee members discussed whether reimbursement should or should not include an egg supplier's lost wages. Given the significant income disparities among women, they considered the possibility that research institutions would naturally flock to "less expensive" egg suppliers, thereby undermining the imperative, as they saw it, that the burden of participation be shared with wealthy women. If they put a reimbursement cap for donors in place that was either too high or too low, they worried that, if it were too high, poor women would be coerced by the financial incentive, but that if it were too low, poor women would be unable to afford to take the necessary time off from work to donate.

> Lansing: We want everyone to have the equal opportunity of any race, religion, nationality to participate in these clinical trials or donation. And we're all agreed on that . . .
>
> King: The problem is equal opportunity to participate. And once you start talking about lost wages and differences and disparities in wages, you fudge together reimbursement [and] compensation. . . . The jury system [in the courts] had to devise a way to cut through differences in income while affording equal protection. I don't know if you want to go there. The way they do it is to reimburse for wages, but at a set fee. . . . It allows poor people to serve as jurors. . . .
>
> Lo: Pat, if I can just paraphrase, your idea eliminates the disparities where one woman gets reimbursed or paid a greater dollar amount than another one with a different kind of job. So it eliminates discrepancy issues. . . . It still sounds too close to paying, buying, selling, which some people object to, but it addresses one set of concerns.[45]

As the discussion continued, several other discussants joined in. Alta Charo is a law and bioethics professor at the University of Wisconsin who serves on the boards of a number of advocacy organizations, including CuresNow and the Juvenile Diabetes Research Foundation. She is also an advisor to the International Society for Stem Research and the Wisconsin Stem Cell Research Program. Notably, Charo also helped draft the National Academy of Sciences' *Guidelines for Embryonic Stem Cell Research* (2005), whose rationale for a universal ban on compensation was quoted earlier in the chapter. Susan Fogel, who was not serving on the Standards Committee, is an attorney and founder of the Pro-Choice Alliance for Responsible Research; she was among a handful of active critics of the stem cell agency who attended many of the committee's initial meetings. Finally, Jonathan Shestack is a movie producer and cofounder of Cure Autism Now, joining the list of board members with Hollywood ties who draw upon their connections for patient advocacy.

> Charo: I feel like I'm completely missing something here. I thought we were talking about lost wages. And suddenly it's turned into something that sounds like payment. If somebody has actually lost $1,000 and you give her back $1,000, she's back to a no gain, no loss situation. Somebody's only lost $10 and you give her back $10, she's back to a no loss, no gain situation. So the disparity doesn't exist once the reimbursement has been completed. There's no disparity at the end if everybody comes out no better off and no worse off than they were. . . .

> Lansing: What I think we're all saying is that [a given research institute's] program, when deciding how much it's going to reimburse—one person for $1,000 and one person for $50—they're probably going to choose the person for $50. In other words, that's the fear. . . .

> Fogel: That the cheapest women will be brought into egg extraction, and that's not what we're trying to accomplish. It's something we're trying to protect against.

> Charo: So part of the group here was worried that the reimbursement limits were needed in order to ensure that poor women were able to donate, but now you're saying that the real problem is that what you're trying to do is ensure that rich women will be donating as well as poor women by making sure that nobody gets reimbursed a whole lot.

> Shestack: It was a public relations issue. Someone who earned a thousand dollars a day gave up five days, and in the annual report it says you paid someone $5,000 as opposed to if you paid somebody $500.[46]

The uniquely public context of CIRM's science-making mandate is particularly pronounced in the way that supplying eggs for research is posited as a kind of civil right which all women should be able to exercise. This inclusionary framing was routinely challenged by Susan

Fogel, whose work as a women's health advocate led her to emphasize the potential risks of economically poor women being overrepresented and taken advantage of; she posited that research institutions might try to stretch their budgets by actively recruiting women with lower incomes. Fogel complicated the bioconstitutional "right to donate" that Lansing was most concerned with, by highlighting the problem of potentially widening the income gap between poor and rich donors. This also brings us back to the issue raised by the Pro-Choice Alliance for Responsible Research earlier in the chapter: far from taking an antiscientific stance, Susan Fogel and others were lamenting a *lack* of scientific rigor that had resulted in insufficient knowledge about the health risks associated with egg harvesting for fertility treatment and stem cell research. They questioned how women could provide "informed consent" to participate when not even the researchers were fully informed about all the possible effects of the procedure.

In the following excerpt, Fogel and board members discuss the competing goals of attracting a diverse pool of egg suppliers without at the same time creating a "market in eggs" to which low-income women will be drawn, to their detriment. Kevin Eggan, Harvard University professor of molecular and cell biology and principal investigator at the Harvard Stem Cell Institute, joins the discussion:

> Fogel: We all have the same goal, which is not to create a market in eggs, not to make low-income women and young women feel coerced by the money, the carrot out there. And what I'd like us not to do is throw a lot of money at women and call it reimbursement. . . .
>
> Lansing: What I'm really concerned about when we get to the end is trying to have a diverse pool. . . . I'm worried that a woman who really wants to do something [that is, donate] and is earning minimum wage will not be able to afford to not work for a day. . . . How are you going to get this diverse population?
>
> Fogel: I don't think we should be confusing race with economics in the first place.

Lansing: I'm not. Let me for the record state that is not what I
meant by diversity, either. You want a diverse pool of people. So
what I'm trying to say is, how are you going to protect somebody who cannot afford to miss a day of work?

Fogel: Perhaps she shouldn't be donating her eggs. Well, I mean,
if she wants to she will, but I'm not sure that creating a whole
package of reimbursement that becomes equated with compensation is just being called something else. I think that's just as
problematic. I think the intent was there will be women who
want to do this. Hopefully, there will be women across the spectrum who will want to do this. . . .

Eggan: As I understand it, there are laws, federal laws, which guide
regulation of human subjects research. And they mandate that
citizens of this country should have an equal opportunity to
participate in human subjects research. And that's what Sherry
means by diversity. . . . And that is why it is important to reimburse people for their participation, because it should be possible
for a woman who is low income who has a relative who suffers
from a debilitating disease who wants to participate in some sort
of research to be able to do that. She should not be denied that
opportunity because she cannot afford to miss a day of work.[47]

The use of "diversity" here is explicitly challenged and then clarified, as
meaning economic and not racial diversity, without speakers ever explaining why it might be important to account for either kind of difference in the donor pool. Rather than go so far as to employ terms
that connote affirmative action, such as "preferential access" or "racial
(or ethnic) quotas" to delineate a desired policy of inclusion, we observe how discussants invoke the sanitized, legally uncontested,[48] and far
more nebulous notion of diversity conveyed, in part, by the inclusionary logic of the NIH Revitalization Act.[49] For a moment, we observe
how Lansing becomes defensive in her invocation of diversity, finding
it necessary to make her reasoning understood "for the record." The
relationship between this discussion and the subsequent debate (below)

over donation as "labor" or "service" is also evident in Fogel's comments that a woman who really "wants" to donate will, regardless of reimbursement. These tensions map onto each other, constituting competing strands of bioethical principle and policy that seek to both protect and empower socially subordinate women as potential egg suppliers.

Labor Versus Service

In the following dialogue, the tension between two framings of supplying eggs as constituting either an altruistic act or a form of work (with extraction of tissue seen as part of a longer chain of biomedical production) is evident. Some discussants express concern that if a donor's normal wages are reimbursed, then poor and rich women will be unequally compensated for the "work" of donating eggs. Marcelina Andaya Feit, president and CEO of Valley Healthcare Systems, is one of ten "disease advocates" on the stem cell board, representing diabetes. In the midst of this consideration of socioeconomic differences, Feit argues that a poor woman can be motivated by feelings of altruism "just like" her rich counterpart. She appears somewhat offended that her peers would assume that women of lower economic status participate in research primarily on account of the financial incentive, and not for the purportedly more noble goal of advancing science and curing illness.

> King: If I were a poor woman who earned minimum wage, and you asked me to be a donor, and when I came in, you said I'm going to give you what your minimum wage is. And then I was sitting in the same room with a woman who had come in and she too is donating for the same reason I'm donating. We have diabetes in my family. I care about what can happen. And I say, how much are you going to get? And she says, "Oh, my daily salary is $500 an hour." That's what I'm going to get in this program. I would say—I can't tell what I would say. It would be, Let me get out of [here] as fast as I can get out of here. . . .
>
> Feit: A poor women who has a family history of diabetes or, let's say, some other, neurological disease wants to donate to help

find a cure for her loved ones. So she's already making a moral
and ethical decision to help her family. Has nothing to do with
her economic status, and [she] wants to be part of the research.
So I think we're selling that individual a little short by all of
a sudden starting to whittle down into some economics. . . .
My daughter, who's a stay-at-home mom, would challenge the
[notion that] just because she doesn't get a paycheck, that she
isn't worth something and hasn't lost something. So we get into
that real slippery slope. I think, first of all, women who want
to do this make that first moral and ethical decision that they
have a reason to want to participate. And I don't think it has to
do with economic status. . . . When we try to circumvent that
initial feeling—I want to be part of the research—regardless of
who they are, then we put in a discriminatory beginning [to]
the whole process.[50]

King and Feit's exchange is characterized by concerns about the subjec-
tive feelings of richer and poorer women, individualizing what are in
effect the outcomes of entrenched social hierarchies. Arguably, broader
equity would first have to be established in order to ensure equity in the
context of stem cell research. In the following conversation, we observe
this tension persist, but here strict adherence to government policy and
deferral to women's health advocacy pressures serve as a proxy for fron-
tal engagement with a social justice agenda. Robert Taylor is professor
of medicine at Emory University.

> Lo: Senator Ortiz has a bill [SB 1260] to forbid that in California,
> you can't compensate for lost wages in oocyte donation. Other
> members of the public have also objected to compensating lost
> wages. And again, this refers to the Pro-Choice Alliance for Re-
> sponsible Research and [the] Center for Genetics and Society,
> among others. They say that we should strike lost wages as a
> permissible expense, and they cite that low-income women are
> less likely to be given permission by their employers, so it ac-
> tually discriminates against low-income women who wouldn't

be eligible for compensation. And also they say that there's a problem with its perverse incentive to get researchers to find low-income people [so as] to stretch their research budgets. . . .

Lansing: I do believe that the woman who has less economic means will be deterred from coming in if she cannot recoup her lost wages, and I think that will be a problem. And I'm more concerned about that inequity than I am about a woman of substantial means being deterred, because she really will say, "I don't need this money" and "I don't want it." . . . I do think in order to protect the low-income woman, we need to do something. I think I like what Jeff said and John said, that we basically say that, you know, we're going to reimburse wages with a cap. . . . therefore somebody who is disadvantaged will not be disadvantaged in donating their eggs.

Taylor: I don't see the benefit of the cap. . . . I'm just kind of concerned about the political fallout of establishing any kind of numerical cap or floor or whatever you decide it is.

Lansing: Well, I'm equally worried that [research institutions] will reject people whose income is too high. Do you know what I mean? This is sort of like an equalizer. It's like jury duty, and I actually think that's a model that serves us well.[51]

We observe how the opposition between the motive to donate in the hope of achieving cures that will benefit one's loved ones (which may be seen as ethically justified) and the motive to donate in order to obtain income that may be of more immediate benefit to those same loved ones (seen as ethically *un*justified) is implicit throughout the deliberation. Discussants' acceptance of this ethical distinction is overdetermined by a culturally salient opposition between (pure) tissue gifts and (profane) biowork. We also observe both the communal and the commodifying potentials of tissue donation at work in discussants' invocation of donors as *either* selfless givers *or* bioworkers, even when these are not mutually exclusive.[52] But in terms of these two ideal types, "women as selfless givers" is compatible with the agency's need to portray research participants as "part of the creation of horizontal, equitable relationships" of solidarity, which everyone has the right, and even the obligation, to join in[53]; whereas "women as bioworkers" bespeaks the hierarchy of most labor/production contexts and comes dangerously close to exposing the vertical, inequitable relationships that characterize the larger social landscape within which stem cell research is being pioneered.

This discord never manifests as explicit conflict in the course of deliberations about how to safely and equitably recruit egg suppliers. Instead, the discussion shifts to concerns about legal compliance and political expedience, with most everyone concerned with the public perception of the stem cell agency's compensation policy. The always present possibility of being charged with not strictly adhering to prohibitions against egg donor compensation (as embodied in SB 1260) proves just as important as the ethical argument against differential compensation based on income—although, as we will see, even this concern wanes in the face of a looming "egg shortage."

Economic Class Versus Biological Race

In the discussion that follows, members of the Standards Committee consider the importance of genetic diversity and whether or not this correlates with racial-ethnic diversity. As before, the role of each discussant is worth noting. In addition to Eggan and Lo, the discussants here were Kenneth Taymor, an attorney with the Stanford Program on Stem Cells in Society; Zach Hall, the former director of the National Institute of Neurological Disorders and Stroke, who served as the stem cell agency's first president (2005–7); Kenneth Olden, senior investigator for the National Institute of Environmental Health Sciences; Geoff Lomax, senior officer for the Medical and Ethical Standards working group, who is trained as a public health practitioner; and Linda Giudice, chair of the Department of Obstetrics, Gynecology, and Reproductive Sciences at the University of California, San Francisco.

Here we can get an idea of the scientific and political motivations for attracting a diverse egg donor pool. More importantly, we can see how scientific uncertainty about whether social categories of difference

map onto scientifically efficacious distinctions for research purposes fails
to deter discussants from proceeding under the assumption that "diverse
cell lines" is always an important aim.

> Taymor: I think that where you are dealing with public relations,
> and to the extent that, as researchers, you feel that the harvest-
> ing of oocytes from a diverse population is necessary in order
> to effectively conduct this research, it would be very benefi-
> cial to the public debate that's ongoing and likely to ensue that
> there be some evidence-based decision-making going on. . . .
>
> Eggan: I think it's important to say that I don't know of any evi-
> dence that suggests, although Ann [Kiessling, director of the
> Bedford Stem Cell Research Institute] could correct me on this,
> that there is a need to get oocytes from a racially diverse group
> of women. . . . We need diversity in the people that are donat-
> ing their eggs. The commitment to diversity is the commitment
> to equal opportunity to participate in human subjects research
> as mandated by federal legislation. That's the responsibility of
> this group, and that's what everyone is trying to live up to. I
> think nothing more and nothing less than that.[54]

This exchange between Taymor and Eggan is one of the only instances
in which the need for a racially diverse egg donor pool was explicitly
questioned, and scientific evidence for such a goal articulated as a necessary prerequisite to actively striving for such a goal. Some analysts have
argued that without a concerted effort, the "differences in biological access to stem cell therapies for different ancestral/ethnic groups are likely
to be large."[55] The use of an ethnoracially homogeneous sample of IVF
donors, in other words, decreases the likelihood that "members of non-
white ancestral/ethnic groups would be able to find matches within
the stem-cell bank," a disparity that seems to mirror the already well-
documented disparity in organ donation in the United States and Europe.[56] Even so, the evidence is lacking; but as we see in the following
dialog, at a Standards Committee meeting five months later, not only
was a presentation of scientific data not on the agenda, but the need for

diverse egg suppliers was institutionally enshrined by making it a feature of the grant evaluation process.

> Hall: One of the issues [in the strategic plan] is how to assure diversity in the cell lines that we derive. And we, in fact, plan on having a special focus group on that issue. . . . I think we are going to take positive steps to make sure we are able to get the kind of diversity I think all of us believe we need going forward if this is to be successful as a therapy. . . .[57]

> Olden: Do we have a requirement in the grant application [from prospective research institutions that seek CIRM money] that the applicant demonstrate to the review committee or the administration of the program that they have in place a plan that will ensure diversity in terms of their sample selection? . . . Because I wouldn't want to leave it up to an investigator unless he or she demonstrated in the application that they had a plan that would really ensure the inclusion of persons from low socio-economic background. If that's in place, then I think we can leave it up to the investigator to decide.

> Lo: [There is some provision for diversity in the standards on grant selection under] "fairness and diversity in research." We say CIRM grantees shall comply with the California Health Research Fairness Act, California Health and Safety Code . . . inclusion of women and minorities in clinical research. . . . I don't know if that specifically says you've got to demonstrate in your grant application that you have a plan to ensure adequate representation. . . . You have to show how you are going to achieve diversity. . . .

> Lomax: What that law says is if you do not already comply with the NIH 1993 rule on diversity, then it essentially mandates the development of something that's substantially similar. So it's pegging it to the NIH policy.[58]

Almost exactly one year following this discussion, the Standards Committee arranged for a formal presentation of health- and risk-related data for egg donors for assisted reproduction treatments. A number of interesting rationales for recruiting different kinds of women for egg donation are evident in the following excerpt. The issue is obtaining data that show the effects of donation on different kinds of bodies, the reasoning being that a combination of socioeconomic status and biological constitution leads to different levels of vulnerability among women.

> Olden: What are we doing to encourage women from socioeconomic disadvantaged backgrounds [to donate]? . . . I guess the women who were in the study [on potential health risks to donors] were mostly upper income, upper middle class or higher. But because I'm thinking . . . diet, for example, may be a very important factor [in risk for Ovarian Hyperstimulation Syndrome]. The immune system, whether one is immunesuppressed, may be an issue. And I assume there are mainly Caucasian women [in the study]. Was that the case? Well, at least the low-income Caucasian women should be looked at, and we should do something, because I remember we spent a fair amount of time discussing that issue to make sure we had a full spectrum of socioeconomic class as well as race be included. Now, grant you, we have to develop some initiatives to encourage these populations to donate eggs, for example. So I think that the fact that we don't have data on low-income Caucasians is a weakness, and we should do something to encourage that, it seems, because these people are likely to be malnourished, and certainly that may have some effect on their risk.[59]

The current data drawn from high-income couples seeking fertility treatment are thought to pertain to healthy, white bodies, even though those seeking fertility treatment may, by definition, not be completely healthy. Nevertheless, it is assumed that their wealth and whiteness confer some protections against the side effects that can result from the hormone manipulation and surgery that egg donation entails. Poor women, who are disproportionately racial-ethnic minorities, are thought to have poor diets and suppressed immune systems that cause them to be more vulnerable to complications. In an effort to determine these negative effects, poor white women emerge as a viable sample to study, though they are often treated as a surrogate for minority women who constitute the actual target population to be recruited as donors.

The discussion then turns to consider how these data limitations might affect the ability of physicians to recruit clinical trial participants. Political expediency is offered as one reason to proceed with the diversity mandate, as researchers are beholden to an "inclusion and difference paradigm" enshrined in the Revitalization Act.[60] A second implied motivation relates to discussants' acceptance of a paradigm of clinical practice known as "racialized medicine"[61]: In order to ensure that stem cell therapies are sufficiently tailored to a racially and ethnically diverse population, researchers must recruit a sample of tissue donors that corresponds demographically to the potential clinical population that will utilize stem cell therapies.

In addition to Patricia King, Bernard Lo, and Robert Taylor, who have been previously introduced, discussants in this excerpt include Ted Peters, a theologian and ethicist and a professor at the Pacific Lutheran Theological Seminary; and Janet Rowley, a professor in the Department of Molecular Genetics and Cell Biology at the University of Chicago.

> King: I mean, there are lots of deficiencies [in the data on egg donation], but the one I'm most interested in is that what we know comes basically from upper-income women, not lower-income women and certainly not racially and ethnically diverse women. And I've been around long enough now . . . to see that when you move into the clinical arena on that kind of base, you never come out of it. . . . What research we have doesn't apply to all potential donors. There are clinical implications of that in terms of how you got informed consent, how you try to do your recruiting, and *why* you try to do your recruiting. . . .
>
> Lo: Let me push a little beyond that. . . . I guess I would tie it back to what Ken Olden said, sort of wanting to have the pool of oocyte donors for research to some extent be representative of the popu-

lation as a whole. And there are scientific reasons for wanting to
do that as well. . . . If I'm talking to a woman of color to whom
the current data on risk may not apply for a number of reasons,
we're trying to say it would be good to have more oocytes from
people like you. On the other hand, in all honesty, we'd have
to say all the data we know about risk is not from women like
you. That strikes me as a very complex discussion to have, but it
would be very important.

King: That is exactly the dilemma that I see, but I think the reasons
for the pool being representative are really critical.

Considering the issue of informed consent once stem cell research has
developed to the stage of clinical trials, Lo imagines a physician trying
to recruit a woman of color and telling her that none of the data on
potential risks of the research were obtained from women "like her."
Yet, unless the physician in this scenario conflates physical appearance
and ancestry with significant genetic variation, he cannot actually know
whether or not the available risk data apply to his patient.[62]

In the following excerpt, we observe one of the only mentions of
racial-ethnic minorities who are *not* presumed to be economically poor;
discussants assume this population of fertility patients share donation
risk factors with prospective minority research egg suppliers. Interestingly, discussants do not make a similar parallel based on economic
status, such that middle-class minorities would be comparable to their
middle-class white counterparts. This illustrates how a genetic prism of
health and illness easily displaces the use of economic class status as a
determinant of a woman's risk for complications.

Taylor: It would be nice to come up with some of the best practices
based on the pragmatism of the most experienced people, maybe
understanding limitations because [the data on Ovarian Hyperstimulation Syndrome risk factors are] quite nonrepresentative,
because typically the couples seeking donor oocytes are of a particular socioeconomic, ethnic sort of slice of the population, and
they're looking for donors that sort of match those characteris-

tics. And I do anticipate that there are going to be significant differences. . . . [I]n preterm labor, for example, we know there are the TNF Alpha Promoter Polymorphisms [a mutation associated with susceptibility to cervical cancer] that probably predispose African American women to preterm labor more than Caucasian women. Those proinflammatory sort of genetic polymorphisms might well predispose to more inflammation and a greater risk for oocyte retrieval as well. . . . The risk of infection and inflammation, at least that we pick up clinically, seems to be quite low in this [IVF] population, in . . . the predominantly white population that's been studied. We honestly don't know what that's going to translate to. . . . It's a puzzling situation . . .

Peters: Then over the long haul, the second related issue is how do social and economic and dietary and racial factors refine what these [egg donation] risks could be. . . . It's up to you as a researcher, if you got CIRM money, to minimize the risks for these people . . .

King: It is not the case that African Americans and other ethnic groups do not use reproductive technologies. They do. They tend to be high income, too. And the question is, is there any way we can catch . . . at least get some of the [assisted reproductive] experience [of these groups when it comes to risks and side effects] where it is available? I don't think you should have any trouble [collecting this data] as long as we think about it in advance and select from certain regions of the country—not just for African Americans here, but for all the racial and ethnic minorities. There are certain places that you can go where you will find a greater impact from certain groups. . . .[63]

Lo: It would be helpful if Geoff and staff would obtain from expert consultants best practice interim guidelines on how to minimize [oocyte donation] risks for minority women and other women who are not represented in the database of current studies by including people—by including among those consultants—people with experience with such populations . . .[64]

Rowley: [A]nd the question is really, as you stated it, it was women who are egg donors who belong to minorities. And I think that the broader issue is, is there a difference in response in women of different ethnic groups or socioeconomic backgrounds across the board—not just egg donors, but across the board? If there's general evidence that one group is more sensitive to whatever sort of treatment, this should be public knowledge. And what we've agreed is that, at least as far as we're concerned . . . as far as an educated public is concerned, there isn't information on different susceptibility or different response of women of different ethnic or social groups to the treatment. And this is what we want CIRM to get information on.[65]

Central to this discussion, as in connection with Lo's fictional clinical encounter, we find discussants concerned with the question of how to cultivate minority donors' trust as a prerequisite to obtaining their consent to participate in clinical trials down the line. But as I argue in the following chapter, the focus on distrust as a psychosocial sentiment overlooks the materially embedded constitution of social relations. Appeals to participate in experimental research, and to supply tissue samples toward the distant goal of advancing regenerative medicine, must contend with the ongoing inaccessibility of basic healthcare for subordinate social groups. Herein lies the strength of the somewhat paradoxical notion that providing donors with a tangible benefit, not simply "reimbursement" for time spent, could perhaps generate greater equity than a policy of noncompensation.[66]

Conclusion

At a 2006 conference on the Ethical Worlds of Stem Cell Medicine, medical historian David Rothman articulated the central concern animating the work of feminists who advocate a reciprocal exchange between egg suppliers and researchers: "What do we owe each other?" This question of debt and obligation is even more crucial in the context of state-sponsored stem cell research, where the ethical treatment of research participants is coupled with the political rights of different demographics of women. Some, like Rothman, are interested in noncommercial alternatives for enacting reciprocity between egg suppliers and researchers. Thus, one commentator argues "that giving someone [as in a donor] some measure of control over the use of a cell line does not necessarily entail that they should also be given an economic interest in the cell line."[67] He disaggregates women's relative control and agency in the egg supply exchange from a commercial model of reciprocity and instead situates it within a social justice framework, wherein women's ability to temporarily commodify parts of their body is within their prerogative.

The broader question about how biological lives and political lives, *bios* and *demos*, are implicated in participatory science brings us to the inclusion of feminist concerns about stem cell research. As with the bio-constitutional struggles discussed in previous chapters, we observe an ongoing tension between political rights and what is *ethically* right, and see this tension amplified by the competing feminist framings of compensation that understand the substance of "rights" differently. Yet despite the centrality of women's bodies to this pioneering field, there is no designated "women's advocate" on the stem cell agency board in the same way that seats are designated for specific disease conditions. Despite, or maybe because of, this lack of formal inclusion, the question of what "women's interests" are and should be in relation to stem cell research continues to be highly contested and, to this point, unresolved.

In the case of California, as I have hinted throughout this discussion, there have actually been two phases in the development of its egg donor compensation policy, corresponding to the two competing feminist schools of thought. In the first phase of setting up the agency, we saw the "institutionalization of donors' self-sacrifice,"[68] wherein compensation was thought to corrupt the process. But as concerns over an egg shortage emerged as a real threat to the progress of stem cell research, the agency amended its strict noncompensation policy to "grandmother in" previously banned tissue (that is, tissue that IVF patients want to donate but whose gametes were originally paid for in the private sector).

Under the initial regulations, even though IVF patients who wanted to donate their embryos to research would not be paid, the fact that one or both of the gametes (egg or sperm) used to produce the embryos had been paid for in the clinical context meant that these "leftovers" were off limits to researchers: if any part of the embryo had been paid for, then it could not be used in California stem cell research. Even though the original sperm and egg donors in the clinical context did not technically have any say over these tissues, since the embryos actually now belonged to the IVF parent(s), the latter could not donate them for research under the original policy.

After three years of near consensus *against* compensation on the part of those implementing the initiative, researchers began to openly decry the scarcity of available tissue under then current regulatory conditions. So at a February 2008 meeting, the California agency's new president, Alan Trounson, created an uproar by announcing that he wanted to reopen the debate about compensation, invoking an ethic of reciprocity in which researchers could obtain an adequate supply of tissue if donors received some tangible benefit in the exchange. The agency now entertained loosening its prohibition in the form of the aforementioned "grandmothering-in" policy, which allowed researchers to use embryos that were produced in the IVF context, where IVF patients paid for the gametes, as long as the transaction occurred *before* the stem cell agency's noncompensation policy had been introduced. The regulatory board playfully introduced feminized nomenclature in place of the conventional "grandfather clause"; I employ it more seriously, to signal the disjunction between feminist-inspired policy discourse and the more (scientifically and economically) instrumentalist conditions that are actually driving the policy shift—the appropriation of feminist politics for other means.

In contrast to California, New York made headlines in 2009 by becoming the first and only state to allow egg donor compensation,[69] to the consternation of many.[70] In so doing, the New York stem cell agency has undertaken a more fundamental reversal of its strict stance against compensation: the New York Stem Cell Science (NYSTEM) board "agreed that it is ethical and appropriate for women donating oocytes for research purposes to be compensated in the same manner as women who donate oocytes for reproductive purposes and for such payments to be reimbursable as an allowable expense under NYSTEM contracts."[71] Payments can be up to $10,000, comparable to clinical compensation, and apply only to egg donation solely for research, not to leftovers from the clinic.

Almost certainly referring to the perceived shortage in California, the New York agency explains that "experiences in other jurisdictions indicate that lack of reasonable compensation to women who donate their oocytes to stem cell research has created a significant impediment to such donation, limiting the progress of stem cell research."[72] The primary reasoning used in New York was that the distinction between egg donation for the purpose of assisted reproduction in the private sector and donation for stem cell research in the public sector was arbitrary and unfounded, since both entailed the same set of risks—and that stem cell research arguably had the potential to have a greater positive impact on society.[73]

The implications of the California stem cell agency's first tightening and then loosening restrictions on egg donation without undertaking as thoroughgoing a reversal as New York are twofold. First, it reveals how the liberal notion of the individual rights-bearing woman can be used in advocacy and in institutional decision-making to both support *and* oppose compensation. When concerns about a shortage grow, the grandmother clause allows eggs to come in "through the backdoor" of the stem cell lab[74]; gametes that were bought, sold, and stored prior to the compensation restrictions are now available to researchers.[75] Publicly funded scientists can benefit from excesses in the private tissue market, with its purportedly overstocked freezers, without making fundamental changes to the logic of exchange in the public domain, where donors continue to be the only part of the scientific production chain that go without compensation.

A second, more insidious implication of the tightening and subsequent loosening of restrictions relates to the stratification and racialization of egg donors—which as we observed was an initial concern in the agency's deliberations. Poor women and women of color were invoked by agency directors and board members as people who needed to be protected as vulnerable subjects that might be unduly coerced by compensation packages but who simultaneously needed to be given an opportunity to donate, for a variety of reasons. The former president of the agency, for example, explained that "assuring diversity in the cell lines" was a prerequisite to success in developing therapies that would be efficacious for a diverse population.[76]

In contrast to this initial preoccupation, however, the issue of diversity is strikingly absent in the agency's current attempt to address the looming egg shortage. There has been no deliberative focus or public comment forum about how socially subordinated women stand to lose or gain as a result of this policy shift. In short, we observe the fickleness of a "people's science" in which feminized rhetoric (the Grandmother Clause) supplants thoroughgoing deliberation regarding the multiple and conflicting ways in which stem cell research impacts socially stratified women. Instead, the needs of stem cell researchers appear to trump the concerns of both of the two feminist camps under consideration.

The result is a complex set of institutional demands that first sought to both protect *and* strategically include a racially and ethnically diverse sample of egg donors, but which now seeks to give researchers access to a private surplus—a policy shift that implicitly offers "vulnerable women" full protection (via exclusion) but also no benefit. Here the ethic of protection conceals an underanalyzed effect of the grandmother clause: if IVF leftovers make up the bulk of the embryos used in stem cell research, then ethnoracial minority women are likely (owing to the racial homogeneity of IVF donors) to receive limited benefit from a "people's science" if and when stem cell therapies are produced. In other words, the "politics of inclusion" ebbs and flows depending on the relative abundance of the tissue market.

CHAPTER FOUR

RACE FOR CURES

Why am I in such demand as a research subject when no one wants me as a patient?

—*Sonny Dixon*[1]

Unless we do outreach to show that these people [sickle cell patients] have been counseled and educated and had access to scientific information to preserve their lives . . . we're going to prejudice the minority population, in the first area of stem cell research they see which might have potential application to them, against participation.

—*Robert Klein, former California Institute*
for Regenerative Medicine chairperson[2]

UNLIKE MORALLY CONTROVERSIAL STEM CELLS derived from human embryos, "adult" stem cells are derived from the specific tissue in the body that they help maintain and repair. Hematopoietic stem cells, for example, are highly concentrated in umbilical cord blood, making cord blood a prime renewable source for remedying blood-based ailments. Sickle cell disease, which predominantly affects African Americans in the United States, is one of the few illnesses for which there currently exist hematopoietic stem cell treatments for tissue regeneration; as such it offers a window into an expanded terrain of stem cell politics characterized by racial/ethnic and class inequality. Here we find a paradox between ethnoracial inclusion in stem cell research and the lack of quality healthcare—too often resulting in preventable, premature death—experienced by many African Americans and increasingly a wider swathe of people in the United States.[3]

In the case of sickle cell disease and its intersection with stem cell research, we observe not only the limits of conventional bioethics as a means to grapple with how entrenched social inequality shapes people's experience of novel treatments, but also the limits of a strictly biological understanding of illness. After all, if the experience of those affected by an inherited single-gene disorder such as sickle cell disease is still very much mediated, aggravated, or ameliorated by intervening socioeconomic factors, then surely less genetically specified illnesses are equally if not more disposed to ultimately socioeconomic influence.

When I began fieldwork at the Garvey Research Complex[4] in early 2005, researchers were utilizing a federal grant to collect and store cord blood units from families across the country to use in stem cell transplantation. Even though Garvey serves a large population of patients who were potentially eligible for a transplant, researchers expressed frustration at the low enrollment rates, especially among the predominantly African American sickle cell population. While they enrolled families affected by leukemia and other rare disorders, their primary target was families with a child diagnosed with a hemoglobinopathy, an inherited single-gene disorder in which there is abnormal production (for example, beta thalassemia) or structure (resulting, for example, in sickle cell disease) of the hemoglobin molecule. The sickle cell disease priority explains why the program could draw on the federal grant to fully subsidize the stem cell transplant procedure for hemoglobinopathy patients but did not typically offer the same to enrollees with other illnesses.

After identifying eligible families (that is, those expecting the birth of an *un*affected child), caseworkers walked parents through the collection of umbilical cord blood, which would then be cryopreserved and released for use in a transplant to the affected older sibling, with the understanding that the odds for a good match are much better with sibling donors than with unrelated donors. By 2007, the program had managed to collect approximately two thousand units of cord blood, five hundred of which were from sickle cell patient families. However, of those siblings who were well matched, only 6 percent of sickle cell patients ever underwent the transplant. Compare this with the *60* percent of eligible beta thalassemia families that consented to the transplant—and who, as it happens, were of predominantly Asian descent—and the ethnoracial contours of this novel treatment start to come into focus.

From the perspective of many proponents of the procedure, African American patients are "underutilizing" stem cell transplantation. Typical explanations for African American resistance to such experimental medical treatments range from psychological disposition (for example, distrust) to the structural marginalization of African Americans (for example, lack of access to health insurance); yet, these are insufficient explanations. By attending to the relationship *between* disposition and socioeconomic position,[5] we see that distrust is socially produced in the everyday experiences of patient families in and outside of the clinic.

For stem cell transplant enthusiasts, African Americans' underutilization of stem cell treatment is especially difficult to grasp, because the 85 percent event-free (that is, with minimal complications) survival rate is considered exceptionally good for a relatively new treatment. The ways in which stem cell program staff account for patient family decision-making are an extension of the differential ethnoracial dynamics they encounter in more routine clinical encounters. To illustrate, consider how Garvey's lead caseworker explained the differential transplantation rates among African Americans and Asians:

"Although it's rather crude, sickle cell patients act like they don't have any control over what happens—fatalistic—and it may be that they don't trust medicine and science. But then thalassemia patients are so controlling. They have a completely different perspective on medicine and science. They absolutely trust it."[6] Here the caseworker invokes popular, racialized notions about science-philia among Asian Americans as contrasted with science-phobia among African Americans—suggesting possible cultural differences among these patient populations to explain the disparity in transplantation rates. This and similar deployments of supposed cultural differences have a long history within social scientific literature. Arising most influentially within the "culture of poverty" framework, the focus on the "learned helplessness" of subordinated groups[7] leads well-intentioned scholars to exhort health providers to "promptly identify fatalistic persons"[8] as the primary locus of intervention in order to increase medical compliance. But not only does this focus, in my judgment, obscure the relative trustworthiness of medical institutions; the focus on fatalism leads analysts to misidentify rejection of biomedical treatments as a lack of agency writ large.

Challenging such cultural generalizations, a number of scholars have examined why and under what conditions African American patients trust or distrust medical professionals[9]; these academics are exploring links between patients' distrust and their unwillingness to participate in biomedical research,[10] drawing on a small but growing body of research that draws attention to the experiences of sickle cell patients specifically.[11] In the majority of this work, medical distrust is operationalized as a set of individually held views—about physicians, medications, specific procedures, and protocols—that are in turn typically measured using survey methods.[12] While this usefully serves to move us away from the cultural generalizations of an earlier era and provides important insights into how differential levels of trust among individuals may impact service provision and health outcomes (potentially deidentifying distrust as specifically a black problem), the focus on individual attitudes does not adequately account for the social production of distrust among U.S. ethnoracial groups. Partly because the thrust of this growing literature aims to hone informed-consent protocols so as to increase human subject participation among "hard-to-reach" populations; create standing trust relationships[13] far in advance of recruitment efforts[14]; and ultimately reduce patient "noncompliance" by improving doctor-patient relationships, this research obscures some of the specifically *social* processes that reproduce estrangement among many patients.

As one young African American man, Sonny Dixon, expressed it, "Why am I in such demand as a research subject when no one wants me as a patient?"[15] On the basis of such comments, I suggest that when routine quality of care is lacking, when healthcare and research are conflated, and when patients feel unprioritized medically, resistance toward the experimental enterprise is a reasonable response. Sonny's query, I contend, leads us into the broader context of medical decision-making that sickle cell patients inhabit, drawing us closer to the contested processes of biomedical recruitment and experimentation, compelling us to contextualize this basic expression of ambivalence and resistance. By reorienting the conventional approach to studying African Americans' "unwillingness to participate" in experimental treatment, then, I draw our attention to the processes that organize people's experiences as both objects *and* agents of medical treatment. In so doing, we are confronted not only with the multiple uncertainties within which scientific and medical decision-making take place,[16] but also with the ways in which Sonny's query extends beyond a biomedical purgatory, a limbo void of reason or action. Rather, as the cases below illustrate, ambivalence toward the promise and risk of novel treatment gives rise to forms of agency and resistance that defy the narrow decision-making frames (for example, to undergo a transplant or not) posited by transplant proponents. We are dealing instead with something we might call "ambivalence in action." That is, people's dispositions arise out of everyday experiences, and hence are mutable and sometimes contradictory rather than reflecting stable, self-contained beliefs about medicine and medical providers. In addition, caregivers (such as parents) are shown to enact nonmedical modes of treatment and care to counter the growing biomedicalization and uncertainty of everyday life with more durable care regimes that seek to prevent and treat the onset of sickle cell–related pain.

To be clear, the narratives that follow are not presented to prove one or another hypothesis about why African American patient families may be reluctant to use stem cell treatments, but rather to take an opportunity to examine what we're calling ambivalence in action. The intent is to leave the terrain unsettled: to show that the ambivalence toward experimental medicine emerges as socially "situated knowledge"[17] that can be both incomplete and true.[18] Such socially enacted ambivalence accounts for the ways in which people's decisions to undergo or decline participation in novel treatments reflect incisive determinations about, for example, how much one is cared for versus how much one is being used to advance institutional agendas. In that way, we are not seeking at this point to predict outcomes but rather to provide a way to conceptualize the connections between shared sentiments and everyday behavior.

Following an important shift in scholarship on sickle cell disease initiated by sociologist Simon Dyson and colleagues, I find that the relatively low sickle cell transplant rate, understood more accurately as ambivalence in action, is fueled by three processes: (1) the therapeutic uncertainties of both novel and more mundane treatment regimes, (2) the institutionalized conflation of healthcare and research, and (3) political struggles in which sickle cell patients were symbolically, but not structurally included in medical investment decisions. I suggest that together, these three aspects of sickle cell patient care give rise to an ambivalent response to stem cell recruitment in which caregivers are justifiably resistant to this latest biomedical cure even as they hold out hope for more robust forms of institutionalized care, and even cure.

Therapeutic Uncertainties

Of the parents I interviewed of sickle cell and thalassemia patients who were enrolled in the Garvey cord blood bank and were eligible for a transplant, the large majority of sickle cell families declined the procedure—whereas over half of beta thalassemia families underwent it. Among the reasons that caregivers repeatedly gave for declining a transplant were that they did not want to use a less established (that is, experimental) procedure for their children, and that they did not perceive that the potential benefits outweighed the known risks.[19]

To illustrate one caregiver's response to what I would call this "therapeutic uncertainty," I examined the case of the Harts, a family who had been utilizing Garvey's services for the previous fifteen years and had developed a keen sense of the various facets of patient care. At the time of my fieldwork, fifty-five-year-old grandmother Sethe Hart was the primary caregiver of fifteen-year-old Destiny Hart. Shadowing the head physician, Dr. Tate Wright, into the hospital room, I found Destiny being checked by a pulmonary specialist, who went on to tell Ms. Hart that Destiny's breathing was a bit abnormal and that she wanted to test whether Destiny would benefit from an inhaler. Hearing that, Sethe Hart proceeded to ask the specialist a battery of questions, closing with, "Why would you give steroids [in the inhaler] to a child?" Lest Ms. Hart's query be seen as misguided skepticism, we should note that this exchange came on the heels of FDA safety warnings about the possible side effects of using the very inhaler brand under discussion.[20] Even so, after further explanation by the specialist, a compromise seemed to emerge: Destiny would use the inhaler every day for a month, and then come back in to see whether it had helped or not.

What I learned a week later, however, when visiting the Harts in their home, was that they had other plans altogether. Turning to the inhaler issue during our interview, Sethe Hart explained,

> Speaking of asthma, we're challenging the test that we're supposed to be in right now. We're not taking that stuff. We're walking [that is, exercising]. So when we go back in a month, they're gonna say, "Oh! It's the results." Check this out. We fixing to make a fool out of them. We're gonna walk every day, build up that breathing, whatever it is they're looking for with them lines on that machine, and we just gonna make a fool out of them. Don't believe everything you hear from man, 'cause if you do you'll be in bad shape, 'cause there're side effects to everything, there're side effects to all medications.[21]

We should note that Ms. Hart expresses concerns about a generalized risk that she attributes to all medications, not just those deemed "experimental" by health providers. This proves to be especially critical for those caregivers whose child has experienced relatively mild symptoms up to that point. Ms. Hart also predicts that Destiny's improved breathing will be falsely attributed by the pulmonary specialist to the inhaler. Though they do not intend to use the prescribed medicine, she advances the claim that a nonmedical method—taking walks to improve lung capacity—is a superior treatment to the inhaler medication. Recalling anthropologist Ian Whitmarsh's "potential asthmatics" who were prescribed medicines as part of the diagnostic process, Ms. Hart is skeptical toward what she perceives to be a tendency to overprescribe medications.[22] Whitmarsh's respondents described pharmacists taking the printed insert explaining serious side effects out of medication containers before giving them to caregivers, "suggestive of a dangerous secrecy"[23] to induce compliance; similarly, several of sociologist Shirley Hill's sickle cell mothers expressed concerns about the administration of penicillin. One of these respondents, who knew that a particular study sought younger children, "refused to participate in the penicillin program by means of waiting until her daughter was technically too old to be part of the study":

> I didn't want them testing her to see if penicillin would help new sickle cell children under the age of five. They had no long-range test of that. And penicillin could block the immune system. What would she do later on when she got older. . . . what happens to your child later on when she can't function because she needs penicillin? . . . I'm sorry, I know you need these experiments and stuff, but this is not the part that we choose to participate in for sickle cell.[24]

Like many caregivers navigating the risks and benefits of agreeing to experimental treatments within the context of their long-term care work, this respondent is not willing to take on the burden of unknown future complications. For some hematologists and their patients, the curative potential of stem cell transplants is not easily outweighed by the 5 percent mortality rate and the potential for serious complications— including, in this case, life-threatening infections, sterility, and chronic host-versus-graft disease, in which the patient's immune system attacks the foreign tissue (a risk minimized but not completely eliminated with close sibling matches). Whereas the vast majority of thalassemia patient families who are considering whether to undergo the procedure are faced with the prospect of a lifetime of monthly blood transfusions and attendant complications without the transplant, the wide spectrum in sickle cell severity[25] makes the decision whether to accept the risks of the procedure a much greater medical gamble for sickle cell families: not only is the transplant outcome unknown, but a patient's disease progression if she does *not* undergo a transplant is also unknown. Some experi-

--Rochester Public Library--
Parking in the ramp is free
if you are in and out
within an hour.

Item ID: 0101815590274
Title: Sigma Xi lecture [videorecording (D
VD)] : acceler
Date due: 3/7/2020,23:59

Item ID: 0101815677642
Title: Supercells [videorecording (DVD)] :
 from stem cel
Date due: 3/7/2020,23:59

Item ID: 0101808962449
Title: The Proteus effect : stem cells and
 their promise
Date due: 3/7/2020,23:59

Item ID: 0101812421515
Title: People's science : bodies and right
s on the stem
Date due: 3/7/2020,23:59

To Renew, call 328-2311 or
online at
www.rplmn.org

ence relatively mild symptoms, with periodic pain crises that families learn to manage, whereas others experience strokes when they are as young as five and require hip replacements by the age of fifteen. The mere diagnosis of sickle cell disease does little to inform parents about where their child might fall on this spectrum, and so willingness to try a high-risk cure is often outweighed by the hope that one's child will be one of the lucky ones with mild symptoms. This personal risk calculus, however, is not sufficient for understanding why sickle cell patients are not undergoing stem cell transplants.

Similarly, I observed Ms. Hart's "ambivalence in action" when confronted with the experimental protocols routinely proposed to her at Garvey:

> Ever since Destiny was born, they have always tried to get me to OK tests, you know, with different medicines. Every time we come to an appointment they want to introduce me and Destiny into a study, and I tell them "No!" every time. Don't even waste your time! 'Cause I don't want them. . . .
>
> When Destiny was young . . . they wanted to do a study to see if she was going to have a heart attack! They were going to inject stuff, give her medicines, and once we leave the hospital, I'm the one who has to give her all the medicines and stuff. And I said, "No way! I'm not gonna do that," because she was still young, and her body was pure, and clean. But the only thing that she had inside her body that was a little defect was the sickle cell, so why go throw something else in the body to be tested, and the body's still pure, clean. . . . This is a little, brand-new body. Don't try to test a brand-new something.
>
> What I've seen with the other patients, the other children, is that because they've done their tests, [their] blood looks funny, [they have] hair loss, they're traumatized. Those kids are a mess. And I believe putting all of those fluids and testing those kids, and the parents have allowed those doctors to do it, . . . is a large contribution to why our kids, our sickle cell kids, are still sick.[26]

Sethe Hart's sabotage of Destiny's prescribed treatment, then, and her rejection of Destiny's participation in medical studies, grows out of her own experience as a caretaker and observer of other children's run-down conditions, which she attributes to their participation in clinical procedures of uncertain therapeutic efficacy.

When I questioned Ms. Hart about the conflict between her frustration with both routine and experimental treatments and her relative support for stem cell research, she answered that while she thinks it's fine for taxpayer money to be used for research that may cure sickle cell disease, she doesn't believe they will ever find a cure, because scientists don't acknowledge the ultimate spiritual source of cures; by seeing themselves as the source of cures, researchers sabotage their own success. Here, Ms. Hart appears to finger scientists as the biomedical saboteurs, pointing to their lack of confidence in spiritual intervention and not her own lack of confidence in experimental protocols as the reason why stem cell treatments may not succeed. What might be regarded as her "distrust" toward medical studies, in other words, could be understood otherwise: as her trust *in* something other than an experimental method. And lest her religiosity appear misguided or even dangerous, consider the growing body of epidemiological findings that show a "protective religious effect on both morbidity and mortality," especially for African Americans,[27] and most notably among sickle cell patients.[28]

Sethe Hart's assessment supports findings from other studies on African American mothers' management of sickle cell disease wherein they draw upon situated knowledge to guide their caregiving practices. Most of their attention is directed towards reducing the frequency and severity of pain crises by carefully monitoring medications, diet, physical activity, and the emotional well-being of their children: "Mothers do not simply follow medical advice; they also learn from experience. Their care strategies are often tailor-made, based on experiences with their own children."[29] These findings suggest a surplus of experiential knowledge that directs caregiving practices and shapes health outcomes. For example, many parents point to a correlation between stress and pain crises, such that they make an effort to reduce stress-inducing encounters (for example, intervening in sibling arguments or withdrawing their child from a hostile school environment).[30] One mother insisted, "A headache can trigger a crisis because it's stress. Stress kicks off sickle cell. Arguments kick off sickle cell. The pain doesn't just start naturally. Something triggers it off."[31]

This mother's understanding about the interaction between environment and biology echoes Sethe Hart's concerns about asthma medications; both resist the dominant framing of sickle cell disease and biomedicine as always and everywhere painful or beneficial, respectively. For Ms. Hart, that ordinary physical activity like walking is "more natural" than the inhaler means it is a superior method of addressing Destiny's breathing irregularities. Not only may the inhaler lead to side effects, but it also involves greater biomedical dependency, which may be understood in terms of time and money insofar as she and Destiny must repeatedly travel to and from the clinic, must wait to be seen each time, and may possibly have to pay a portion of the medication costs. Thus, what is "natural" about exercise can be understood as not only avoiding a medication that may or may not work but also avoiding the expenditure of money and time required to come back to the clinic. Exercise, by contrast, is seen as an activity they would integrate into their ordinary day: a time set aside that grandmother and granddaughter would enjoy spending together, thereby enhancing one another's well-being.

Likewise, sociologist Shirley Hill's respondent describes how stress induces pain crises, and considers a "purely biological" explanation inferior to one that takes into consideration "triggers" of pain aside from red blood cells. For her, a biologically determined framing undercuts her own ability to manage her daughter's condition, while for Ms. Hart an appeal to a "natural" treatment enhances her agency vis-à-vis biomedicine.[32] Most important for this discussion is that caretakers are shown to enact caregiving strategies with or against a "natural" idiom of illness by questioning the inevitability of sickle cell pain and biomedical balm,[33] responding to the broader context of therapeutic uncertainty. But even noting caregivers' personal agency, we observe a similar "ambivalence in action" in both narratives that is directed toward biomedical authority and, I argue, cannot be explained only with reference to their children's precarious disease progression or the unknown outcome of experimental treatments.

Conflating Healthcare and Research

Ambivalence toward experimental treatments is also produced by the institutionalized tension between healthcare and health research. These often give rise to competing agendas that become conflated in the kinds of medical institutions in which the majority of African Americans must seek healthcare,[34] wherein not even the fundamentals of healthcare are available. One father of four articulates this best: "There's an old proverb that [says] when your house is on fire, you don't worry about broken windows. In a lot of communities where sickle cell is present, they're struggling. People are very poor and they have a lot of problems, so they don't look at this particular problem as being one that's overwhelming in relation to other problems that they have."[35] The words of this caregiver bring us back to the social dissonance produced by an overinvestment in experimental research when basic-quality access to healthcare is hard to come by for many people living in a country without comprehensive national healthcare. For many of my respondents, focusing on the 6 percent sickle cell versus 60 percent thalassemia transplant disparity at Garvey is comparable to sweeping up broken glass while the more pressing flames in their lives are left to wreak havoc. Research administrators and proponents of novel scientific initiatives alike would do well to heed this father's structural metaphor to the extent that it forces us to pay attention to the social processes that spawn ambivalence toward novel treatments.

The depictions of sickle cell patients as either excessively stoic or chronically drug-seeking are both racialized depictions that together form the collective experience of this patient community.[36] As anthropologist Carolyn Rouse explains, "When sickle cell patients enter the emergency room asking for strong opioids, the overwhelming response by physicians is to view them as drug-seeking and difficult to manage both medically and socially. This perception is so pervasive that many hospitals try to dissuade sickle cell patients from seeking care in their facilities and refuse to establish affiliated sickle cell clinics."[37] Together, the relative invisibility of pain and the difficulty in establishing a standard of care have led to neglect and mistreatment for many patients,[38] as in the following description by Roxanne:

> Most recently—not here, at another hospital—I went through the emergency room, and there were people coming in with scrapes on their knee and nosebleeds and things like that. And they would go right in. I waited there for eight hours in an emergency room waiting to get treated when everyone else just walked in and out. The doctors didn't know much about sickle cell; they were really insensitive and rude and kind of just brushing me off. [The doctor] said some mean things, too. I didn't get treated for another eight hours. It's really difficult sometimes. I don't even like to talk about it [crying].[39]

Roxanne's and other patients' experience of basic healthcare is a crucial element in the collective ambivalence we have encountered towards stem cell transplants. In at least one hospital at which I triangulate my observations at Garvey, nursing staff have developed a "behavior contract" to curb what they experience as disruptive behavior on the part of some patients: "a zero tolerance policy with regard to abusive/threatening/intimidating behavior. Failure to comply with the following rules will result in your immediate discharge from the hospital and/or the intervention of law enforcement personnel."[40] A white nurse in this teaching hospital intimated that while the contract itself does not specifically name sickle cell patients as the target population, her experience is that the contract was developed for and is selectively applied to such patients. The exceptional penalization of black patients described here is a pronounced feature of African Americans' experiences, not only in the case of sickle cell care but in many other medical encounters as well.[41]

Sickle cell families' decision-making is shaped not only by an assessment of the relative efficacy of a novel procedure vis-à-vis the relative

Race for Cures   126

severity of a particular patient's condition but by the broader politics of access to quality health services. When families experience a disproportionate emphasis on their value to research and much less attention paid to their everyday healthcare needs, skepticism toward research solicitations appears justified. But as a result, appeals to participate in experimental procedures are all too often presented as altruistic gestures in the name of cutting-edge healthcare, where "cutting-edge" can be understood as a euphemism for risk and uncertainty as much as it may point to possibilities of remarkable cures.[42]

Consider, for example, that while Garvey's transplant program is "public" in the sense of being subsidized by federal and state grants, the cord blood units collected through the program are not publicly available. While only families who bank a unit have access to their own unit, the units are available for scientific research once a family no longer needs it (because the ill child either dies or gets a successful bone marrow transplant, or the five-year storage life of the unit draws to a close). Nancy Somers, the lead caseworker at the transplant program, explains:

NS: Well, that's a huge challenge for the program. There's all this blood that we can't throw away.

RB: So what's done with all the blood that the families aren't going to use?

NS: For the affected blood, we used to get lots of requests from scientists who wanted to use it for research.

RB: And you all can do that?

NS: Families have the option of donating it for research once they don't need it. There's a checkbox on the informed consent form when they enroll.

RB: And how many enrollees actually do that?

NS: About 20 to 25 percent check "No," that their collections can't be used.

RB: Oh, so the majority don't mind?

NS: Yeah, they figure that if their family isn't going to use it, the blood may as well be put to use to help *someone*.[43]

Somers's account corresponds in part to the high levels of public support in the United States for innovative research programs that promise to improve the health of the citizenry (for example, the passing of the California initiative), with strong precedents in the U.S. culture of blood and organ donation.[44] From the perspective of racialized groups who have routinely been excluded from or stereotypically derided by mainstream American institutions, however, the Garvey arrangement may very well be understood as a case of purported altruism masking research interests. The same medical institution offering free stem cell banking through government funding makes its unused cord blood available to researchers—a dynamic that is aided by the extremely low rates of blood usage by the families themselves. Consider an excerpt from an outgoing letter addressed to each enrolled family's health provider that seeks to recruit the most well-matched patient-sibling duos to take part in transplantation:

> Recently, we confirmed that a cryo-preserved HLA-identical [histo-compatible] sibling donor cord blood unit collected for your patient, [name inserted here], who has sickle cell disease, has characteristics that we believe makes it suitable for transplantation. Thus, we are writing to inform you about our prospective, government-supported umbilical cord blood transplantation study entitled [name inserted here], for which your patient may be eligible. . . . We hope that you will consider enrolling this patient in this clinical trial.

This correspondence is one of the few instances in which the experimental nature of the transplant "study" is plainly expressed, albeit still within the framework of biomedical outreach to a hard-to-reach population.

This window into stem cell recruitment raises a question about the superiority of "public" stem cell banking as touted by its proponents, emphasizing as they do the greater possibility for reducing disparities in access to biotechnological therapies when blood containing stem cells is publicly maintained.[45] But this ignores the cost to *particular* publics of being so available to research solicitations such as the one above. This cost is incurred because of the purposeful conflation of research and treatment, especially when the rate of successful transplantation is perceived as "low" and the procedure is therefore deemed "risky." Under such conditions, the already fragile relationship between many African Americans and medical institutions may corrode further. By contrast, when the transplant success rate is perceived as relatively high, researchers may justifiably gloss the experimental nature of the process, leaving them open to backlash if and when serious posttransplant complications arise.

Political Contests over Medical Investment

As early as the 1950s, sickle cell disease had become "a new source of income, a commodity in a growing [medical] service sector."[46] Research on sickle cell was characterized by an influx of philanthropic and government funds, so much so that experts in molecular biology, clinical specialists in hematology, patients' rights advocates who sought recognition for this "orphan" disease, and politicians who were determined to lobby for grants on behalf of their African American constituencies all clamored to increase the visibility of sickle cell. At the same time, however, reflecting changing economic and social relations on the national scene, sickle cell disease's newfound celebrity also met with a backlash by those who "simply resented the fact that political pressure from African Americans and liberal politicians had influenced the direction of National Institutes of Health (NIH) research dollars."[47]

In the clinical context, a succession of therapeutic "breakthroughs" were celebrated, then disparaged as serious side effects came to light. Bone marrow transplantation (BMT) was the immediate precursor to the cord blood transplantation procedure considered here, and in many cases it is still used in conjunction with cord blood to minimize immune rejection or to supplant low stem cell counts in the cord blood. Beginning in 1984, BMT was lauded in the *New York Times* as a "life-saving therapy," even though, it was acknowledged, it was "fatal about 30 percent of the time."[48] One of the main factors preventing people from undergoing BMT, however, is the requirement that they obtain a donor with matching bone marrow, and it is precisely this barrier that sibling cord blood collection and transplants aim to get around, since finding an HLA match between siblings is significantly more likely than if one is seeking an unrelated donor. Many of the contentious issues revolving around bone marrow transplantation—"the question of what physicians owed to their patients, what risks and choices ought to be offered to them, and whether BMT should be understood as an 'experiment' or as an 'innovative therapy'"[49]—have also been raised in connection with stem cell transplantation, and there is a very wide spectrum in people's responses, in large part shaped by racial and class politics similar to those encountered in previous examinations of treatment hope and hype.

Against this historical backdrop, it is useful to consider that sickle cell disease is facilitating the emergence of stem cell research as the latest iteration of profitable and promise-filled science.[50] The California Stem Cell Initiative is the largest single government investment in the new field to date, and it includes the building of an entirely new research infrastructure (comprising facilities, training, grant competitions, and the like) to support the research investment. But the public benefits of this taxpayer-funded initiative continue to be called into question,[51] reinforcing the imperative to appear responsive to the state's diverse constituency.

Proposition 71 reveals that the initiative was infused with populist appeal precisely by adducing sickle cell disease as the paradigmatic neglected disease of a potentially underserved public (African Americans). In 2004, when California voters decided whether to invest $3 billion in the Stem Cell Research and Cures Initiative (Prop. 71), both the Yes and No on 71 campaigns made their case with reference to whether the initiative would benefit the African American sickle cell community, for whom adult stem cell transplants were already underway—thereby presumably demonstrating the public benefits of the new therapy. In this way, the sickle cell controversy has elicited considerable political dexterity, as advocates on multiple sides of the debate have drawn upon its racial implications to make competing claims about the initiative's benefits or harms. Consider the official ballot summary produced by the California Attorney General's office, in which the main point of contention is presented as corporate economic (biotech) interests vis-à-vis the will of the people. In the "No on 71" summary, under the heading "Bad Medicine," the sole example provided to support the claim that Prop. 71 amounts to corporate fraud refers to adult stem cell transplantation that uses cord blood to treat sickle cell disease. Among the signatories, oncologist and bioethicist H. Rex Greene expresses his disdain for Prop. 71 in these terms:

> The big question is we live in a community, and the community has to make difficult judgments about how to spend its resources, and if we're actually gonna take resources away from sick people who happen to be poor, who happen to be African American or Hispanic, in the hopes of curing something twenty years after they're dead, that kind of discussion belongs in the legislature, where interests are fairly heard and competing interests have a chance to make their case. What instead has happened, and this by clear intent of the proponents of this initiative, [is that] they went the initiative route because this is the best way to bedazzle and befuddle the public, and they have stayed on message throughout. And they have fooled a lot of really reputable organizations.[52]

Greene—unlike other signatories, whose main concern is fiscal conservatism—identifies himself as a "progressive" who opposes Prop. 71 in the name of "voiceless" racial-ethnic minorities and poor people, who he contends will not be served by this state investment. Though he does not name them explicitly, surely Greene would count the Sickle Cell Disease Foundation among those organizations taken in by the promise of therapeutic gold, made all the more troubling by the omission of sickle cell disease representatives from the stem cell agency's governing board.

In the "Yes on 71" official rebuttal statement, supporters are quick to clarify that the initiative does in fact fund adult and cord blood stem cell research (including that which is used to treat sickle cell disease). This effort to align the "Yes on 71" campaign with the state's diverse racial-ethnic public was vital, because at the time the state's liberal base was beginning to publicly express nervousness about whether the initiative was an elitist endeavor fueled by biotech companies at the expense of women and people of color.[53] While this science "for the people" framing proved successful in the end, the point here is that both proponents and opponents invoked sickle cell as a political proxy that implicitly drew marginalized publics into the debate, if only rhetorically.

As we have noted above, among the patient advocacy organizations that signed on to the "Yes on 71" campaign, the Sickle Cell Disease Foundation was one of only two that were not subsequently represented among the disease advocacy seats on the stem cell agency's governing board.[54] The rhetorical *in*clusion and administrative *ex*clusion of sickle cell disease fuels ambivalence among those affected by an illness that was relatively prominent in the stem cell initiative campaign. This disjunction in turn has material consequences with respect to the agency's allocation of research grants, which may in fact prove a liability for the initiative and may provide a reason to rethink the parameters of public participation in future endeavors.

In the first round of grant allocation, the new stem cell agency awarded over $50 million to research programs across the state but failed to fund a proposal by the only institution primarily serving sickle cell patients. In response, the director of this institution rallied racial justice advocates to undertake a letter-writing campaign to the agency. One of the most important of these letters was from the Greenlining Institute, a "multi-ethnic public policy think tank," that at one point asserted that the state stem cell agency must be held accountable for providing benefits to its diverse ethnoracial public in the form of work contracts and cures for ailments that disproportionately affect people of color. One Greenlining spokesperson expressed the organization's "disappoint[ment] that the application submitted by [the sickle cell–serving institute] was not approved. . . . As advocates for minority health and the elimination of health disparities, we do not believe that the working group appreciated that a proportion of CIRM funds provided by the vote of citizens of this State should be used to support programs that address the needs of underserved communities."[55] As with the other appeal letters, this one specifically linked investment in sickle cell disease with a commitment to the principles of diversity, justice, and inclusion. But the stem cell agency proved unresponsive to these political appeals that focused on health "rights" as being of "collective benefit" for subordinated ethnoracial communities. Rather, Proposition 71 had explicitly codified individual stem cell scientists' "right to research"[56] but did not at the same time institutionalize a corresponding right on the part of social collectives (like the sickle cell community) to shape the direction or access the fruits of this state investment through a process of power-sharing across a wide social spectrum.[57]

After the letter-writing campaign was shown to have had little effect, the sickle cell grant applicant attended a stem cell agency board meeting to make his appeal in person. In a heated debate over the accountability of the agency to California's diverse racial-ethnic population and whether such lobbying undermines the integrity of peer review, the agency's governing board voted ten to five not to revisit the grant decision. Even so, this series of events won several key supporters among the agency's board in favor of the sickle cell appeal, most significantly the HIV/AIDS disease advocate Jeff Sheehy. Following an admonition by a Greenlining health program director that "funding the facilities grant would have been an important step in building a trust relationship with ethnic communities," Sheehy called the failure to fund the grant a "missed opportunity," adding that "these communities would be needed when research moved into clinical trials, and without a prior trust relationship with those implementing the Initiative, it would be difficult to recruit a diverse [tissue] donor pool."[58] Sheehy compared this dynamic to his experience working with HIV/AIDS researchers, who were forced by those affected by the illness to recognize the importance of making good-faith efforts to address the needs and interests of the patient community from which human subjects would be needed.

In short, several prominent stem cell initiative stakeholders argued that funding a grant proposal from an institution with a known commitment to sickle cell disease research could have been an effective preemptive strategy to build political support among the agency's diverse constituency, thereby priming patients' willingness to participate in research. Their warnings reveal that scientists and board members alike are cognizant of the way in which the political exclusion of the sickle cell patient community, both on the governing board and among grant-ees, potentially alienates this population in the clinical sphere.

Conclusion

By focusing on stem cell transplant procedures using cord blood—a more established subfield of stem cell research as compared with embryonic or induced pluripotent stem cells—we are able to examine the complex relationship between medical inclusion and social marginality just as both grow increasingly relevant to the "biotechnical embrace" of contemporary medicine.[59] While the stem cell debate in the United States has largely focused on the moral status of the embryo, the present discussion seeks to contribute to an effort to expand the terrain of bio-ethics as one structured by existing social fault lines; it does so in a way that foregrounds the bioconstitutional struggle at the nexus of sickle cell treatment and stem cell research. By drawing together three explana-tory strands—therapeutic uncertainty, the conflation of healthcare and research, and political contests over medical investments—I am suggest-ing that socially produced ambivalence is an alternative explanation to notions of "fatalism," "personal distrust," "noncompliance," and "scien-tific illiteracy" for why sickle cell patient families resist participating in stem cell research.

Finally, I suggest with regard to future science initiatives that it is better not to conceal, through hyperbolic rhetoric seeking to attract popular support, the multiple uncertainties that characterize public investment in new fields. Rather, by acknowledging the "foundational instability"[60] that characterizes the institutionalization of biomedicine, we may be more prepared to deliberate across social, ethical, and po-litical differences. Providing an important perspective for reimagining healthcare, the convergence of sickle cell and stem cell research impels us to situate the relative burden of caring for an ill child within broader contexts of suffering and resilience which temper the notion of illness as strictly biological. The multiple uncertainties associated with sickle cell disease progression (mild to severe), transplant outcome (event-free or

with severe complications), and socioeconomic context (access to affordable quality healthcare or not) form an unstable nexus for establishing
what is ethically right and what should be sociopolitical rights for this
patient community. Examining the struggle over these questions in context rather than measuring people's discrete attitudes about treatment
via surveys, we can see how caregivers exercise agency as subjects, not
merely objects, of science: reasoning, negotiating, and acting without
great certainty, perhaps, but in ways that strategically reinforce regimes
of long-term care. Likewise, as with other bioconstitutional struggles,
depending on *who* one imagines them to be—biomedical consumers
anxiously awaiting the best new treatment, or collectives that have a
strained relationship with scientific and medical institutions—*how* one
seeks to implement a "people's science" alters accordingly.

CHAPTER FIVE

DEPATHOLOGIZING DISTRUST

The problem of distrusting citizens should be recast or reformulated as an
issue of social justice.
—John Johnson, Andrew Melnikov, "The Wisdom of Distrust"[1]

The organization that owns trust, owns its marketplace.
—Julie Brownlie, Alexandra Greene, Alexandra Howson,
Researching Trust and Health[2]

ON MARCH 14, 2008, a celebrated African American artist, Casper Banjo,
walked from his East Oakland home to the nearby police station to seek
help. Observing what later turned out to be a replica gun, police shot
the seventy-one-year-old Banjo to death. As his good friend, fellow disability activist and artist Leroy Moore, would later explain, Casper's life
chances were determined far less by any biological defect than through
a complex intersection of race, gender, class, and disability—that is, by
the *social* disorder that surrounded him.[3] In the days leading up to the
passage of Proposition 71, this understanding led Leroy to ask whether
the California Stem Cell Initiative "would reach his people and other
people of color who are wheelchair users because of police brutality?"
Then, following the murder of Casper Banjo, Leroy wrote a mournful
tribute in the same online venue, *Poor Magazine*, where he had first
questioned the impact of Prop. 71:

> Casper was a talented and peaceful black disabled artist who
> touched the world with his printmaking, brick layering, black activ-
> ism and . . . his love of stories. . . . Casper Banjo was a great story-

teller. . . . however these stories can no longer be heard in Casper's voice. We, family, friends, artists and members of the National Minorities with Disabilities Coalition, will make sure Casper's story and the night his life came to an end will be told.[4]

Indeed, the act of storytelling can open up civic discourse to the experiences of those who struggle to press the levers of power to address the life-and-death issues that keep them up at night. While the everyday policing and penalization of inner-city black life is well documented in the scholarly literature,[5] the set of injustices related to Casper's murder is less well understood.[6] In March 2010, Los Angeles police shot an African American autistic man, Steven Eugene Washington, twenty-seven, in the head because he was not responding to their commands and seemed to be reaching for a weapon. Like Casper Banjo, Steven Washington was not armed. Only months after Steven's fatal shooting, police in Seattle shot John T. Williams, a Native American woodcarver who was partially deaf, four times because he did not respond to the officer's repeated direction to put a knife, which was later found to be in a closed position, down. What do we make of the growing body count of people who are victimized, in part because of their disabilities, at the hands of those whom we entrust to protect and serve society? How, if at all, do people's experiences of everyday policing relate to their trust in other social institutions?

Social disorder, which leaves people like Casper, Eugene, and John vulnerable to overzealous policing (itself a symptom of poor government priorities, which defund educational and employment opportunities while building prisons), is a cornerstone of the larger context in which individuals, families, and entire communities distrust a government that condones such behavior. In a large 2011 survey of community-dwelling adults, researchers measured the level of what they called *interpersonal* distrust that African Americans and whites have in clinical research versus their level of broader *societal* distrust. They explain that the former is "based on personal experiences and interactions of individuals within health care or clinical research settings. In contrast, societal distrust is characterized by a global negative outlook on clinical research based on perceptions of collective research entities or life experiences in society at large."[7]

In comparing the two types of distrust along racial lines, the researchers considered the possibility that people could hold contrasting beliefs: for example, they could trust their doctor but refuse to participate in clinical trials, much like Sethe Hart in Chapter 4. In fact, the study found no racial differences in interpersonal distrust in clinical research but concluded that African Americans "were more likely to express societal distrust in clinical research." Given that most measures of trust in the medical arena focus on doctor-patient relationships and other interpersonal experiences, the study concludes that "maintaining trust in the patient-physician relationships may be necessary but not sufficient to minimize societal distrust in clinical research."[8]

It is one thing to accept that stratified social groups perceive and experience the state and its various institutions so differently, but we must also account for fundamental transformations that *deepen* societal distrust. In the United States, this means attending to both socioeconomic and sociocultural forms of dispossession[9]; in the medical domain, the valiant attempt to implement "culturally competent" care in response to longstanding cultural dispossession can be perceived as disingenuous unless practical attention is given to socioeconomic barriers to quality healthcare. While poor people have always been stigmatized in the United States as lacking in character and morals, the vehement racialization and concentration of poverty in urban slums in the second part of the twentieth century intensified the preexisting denial of dignity. Sociologist Loïc Wacquant analyzes the new forms of social insecurity produced in "the historic shift from the Keynesian state of the 1950s to the neo-Darwinist state of the *fin de siècle*, practicing economic liberalism at the top and punitive paternalism at the bottom."[10] Wacquant's characterization of society as a survival of the fittest in which poor people are systematically disadvantaged—victims of what he calls the "(de)formation of the postindustrial proletariat"—brings us back to the specific forms of vulnerability and violence experienced by Casper Banjo, Steven

Washington, and John Williams. That is, members of subordinated social groups are not only physically assaulted through police profiling and brutality but also dis-abled in other arenas.

The Wisdom of Distrust

Perhaps, then, distrust on the part of the dispossessed is a rational response to, and a defense against, a society that justifies penalizing the poor so that everyone else can feel safe and secure. When life is lived under a state of physical and symbolic siege—as when police helicopters rumble over ghetto dwellers at routine intervals looking for suspects, or when sickle cell patients seeking pain relief are suspected of being drug seekers—it is little wonder that widespread distrust persists. It would perhaps be more curious to find people expressing *trust* in social institutions, including science and medicine, under such conditions. In a provocative essay entitled "The Wisdom of Distrust," John Johnson and Andrew Melnikov examine how both trust and distrust are learned dispositions that reflect objective socioeconomic and cultural-political conditions. Drawing upon analysis in a different national context—although one characterized by institutionalized inequalities similar to those found in the United States—they explain how "it is important to distrust the lofty proclamations of politicians and other institutional leaders, to cut beneath the political rhetoric to discover who is doing what to whom, and where the trails of money and profit lead."[11]

Applying this understanding to the arena of medical research, Wasserman and colleagues examine the "production of knowledge and distrust of medicine among African Americans" and argue that the framing of distrust as interpersonally rather than structurally produced reinforces the unwillingness of those who already have reason to distrust medicine to participate in research. Referring to historic medical abuses such as the Tuskegee syphilis trials, they argue that

> victimisation of African Americans by professional medicine was largely a pervasive outgrowth of structure, not the idiosyncratic work of particular individuals. Ethical shortcomings of the past were short-

comings of medicine as a whole, not of particular players. Poverty, racism and segregation created a structural context in which acute events could transpire. . . . Ignoring the context of medical ethics has negative consequences by creating fear of doctors and clinicians rather than an understanding of the social contexts in which they operate. This feeds distrust of medicine because a failure to understand the contextual nature of ethics is naturally coupled with a lack of understanding about the progress that has been made in institutionalizing medicine and medical ethics since these past events occurred.[12]

But these authors' optimism about the structural changes that medicine has undergone since such historical abuses transpired, leading to their thesis that if we focus on structure then people's distrust will wane, may be premature. In terms of institutionalizing ethical review of research, surely much progress has been made. But this hardly addresses the underlying inequalities that pervade the delivery and quality of medicine, much less the wider lived experience (racism and segregation) that Wasserman and colleagues outline and that aggravate societal distrust. In other words, they are right—we need to attend much more to structural inequality and not simply interpersonal interactions—but doing so only casts a bigger shadow over the prospects of cultivating trust in medical research.

Despite the possible wisdom of distrust, a growing medical and bioethics literature continues to survey, measure, dissect, and explicate this phenomenon as if it were an inexplicable curiosity. Such analytical summersaults are a reminder that with most knowledge domains structured by specialization, analysts are too often trained to ignore the elephant in the room. We routinely treat problems arising out of structural inequalities as just too big, requiring so fundamental a restructuring of institutions, norms, and values that we opt instead to investigate the symptoms of social disorder, such as distrust, in isolation.

For example, the relatively new specialty that focuses on recruiting "hard to reach" populations for biomedical research is what sociologist Steven Epstein dubs "recruitmentology." Techniques include mass mailings; media campaigns; referrals; community outreach in churches, beauty shops, and barbershops; support groups; health fairs; and video-tapes.[13] The success of research initiatives depends on how "participant-friendly"[14] research protocols are, which in turn depends on substantial "pre-program planning" in order to establish "community acceptance," so that "community leaders and members have a vested interest" in the initiative.[15] This community-based approach to cultivating trust and consent has become more central to the stem cell enterprise as California's stem cell agency has shifted from a strict focus on basic research in its first few years to more of a focus on clinical results.[16] You will recall how Phyllis Preciado had a hard time putting community engagement on the ICOC's agenda. But in light of the coming state elections that will decide whether to renew the agency's public mandate, and in a context where the Obama administration supports the field, the agency has had to be more explicit about representing the state's ethnoracial diversity in a way that corresponds to federal research policy.

While clinicians have long relied on racial-ethnic categories in healthcare administration and treatment, the institutionalization of U.S. Census categories in the research context is a more recent phenomenon. Following the 1993 NIH Revitalization Act, all federally funded research must report the racial-ethnic breakdown of the study population.[17] This requirement followed charges by minority and women's health advocates that studies conducted on white men were not necessarily generalizable to other groups. The Revitalization Act in turn spurred researchers' increased attention to recruiting a diverse study population, not only in federally funded studies but also for state-funded research that sought to have national reach. CIRM was no exception, and it institutionalized similar reporting standards in its grant policy.[18] But these inclusionary policies have themselves come under attack for reinforcing the idea that men and women and different ethnoracial groups are biologically, even genetically, so distinct as to warrant representation in this way. Critics say that the targeted recruitment of ethnoracial minorities for clinical trials is part of a larger dynamic of "racial profiling in medicine" that has a number of problematic side effects.[19] Within this fraught arena, distrust has become a major roadblock, insofar as the very groups which federal law requires that researchers include have remained elusive to the gaze of research.

Outreach or Profiling?

On February 26, 2010, CIRM hosted a Diversity Workshop at the Charles Drew University of Medicine and Science, a historically black institution founded in the Watts neighborhood of Los Angeles, in response to the urban rebellion in 1965, with a mission to serve the predominantly black and (now) Hispanic residents of the area. The goals of the workshop were twofold: to "gain a greater understanding of how population diversity affects, benefits and advances CIRM's mission" and "to use this knowledge to ensure that CIRM's funding initiatives support diversity in regenerative medicine." During this forum, Dr. Maria Pallavicini explained how the University of California, Merced, had to "educate the population of this historically underserved population in the San Joaquin Valley about the nature and value of research. This has been a challenge in a region with relatively high rates of poverty and low levels of educational achievement." She described the university-sponsored flea market booths that enabled students to speak with passersby about research efforts and programs in a way that was, in the words of the report, "meaningful and relevant."[20] The report notes that Dr. Keith Norris of Drew University "highlighted the importance of having research scientists engage with doctors and clinical researchers. The goal of this interaction should be to relate research to broader community health concerns."[21]

Conforming to a deficit framing of patient compliance and participation (that is, a focus on what people *lack*), Norris also observed that "functional illiteracy" (48% of U.S. adults cannot fill out a job application) limits people's ability to participate in research initiatives. But he also mentioned other socioeconomic factors, such as "concerns about the time and expenses (travel, child care, lost income)" that are required to participate in research studies. Accordingly, Norris suggested that "smaller mission based and/or minority-serving institutions," like Drew University, are a potential resource for increasing minority participation in research.

On the basic research side, workshop participants emphasized the importance of developing a "diverse stock of cells" to ensure immune tolerability across the diverse population who will seek access to cell-based therapies in the future. Dr. Louise Laurent of UC San Diego presented, in the words of the report, "results from genetic analysis indicating there is restricted genetic diversity in established human embryonic stem cell lines." The UC San Diego team, notes the report, "is currently developing a genetically diverse collection of human iPS cell lines. The success of this effort depends, in part, on the ability to recruit a genetically diverse group of donors to participate in the project." But what connection exists between community outreach to ethnoracial minorities and the genetic diversity which Laurent is concerned about? Soon after the passage of Prop. 71, analysts interested in "considerations of justice in stem cell research and therapy" warned that a lack of ethnoracial diversity would necessarily result in a lack of genetic diversity and the subsequent exclusion of groups who were not well represented in basic research and tissue biobanks. One influential Hastings Center report argued that "[u]nless the problem of biological access is carefully addressed, an American stem cell bank may end up benefiting primarily white Americans, to the relative exclusion of the rest of the population."

Stem cell therapies that are not derived from patients' own cells (as with iPSCs), it is thought, face the problem of immune rejection. That is, if the proteins that are on the cell surface are not well matched (histocompatible) between patient and donor through a process called HLA-typing, then a patient's immune system could attack the transplanted tissue and lead to serious health risks. According to the Hastings report, "An individual's HLA type is linked to her ancestry"—for which many researchers use race/ethnicity as a proxy. So, the argument goes, for publicly funded research especially, an inclusive policy must be developed so "that all ethnic groups in the general population are in some way equally represented in the bank so that no group has a disproportionate advantage."[22]

But the conflation of ethnoracial and genetic variability is not without criticism. Many social scientists warn that using self-identified race as a proxy for genetic variability entails a number of assumptions and risks. "Medical racial profiling," as it has come to be known, is still being vigorously debated, even as community outreach and medical recruitment continue to be premised on a correspondence between U.S. Census categories, genetic variability, and now with stem cell banks, histocompatibility. But while the appropriation of the "profiling" idiom from police work runs the risk of conflating two different modes of targeting individuals on the basis of their group membership—one seeking to penalize, the other to revitalize—these modes share an underlying essentialist logic which assumes that outward characteristics (for example, skin color) can tell us something about underlying characteristics, whether it is a propensity for criminal activity or predisposition toward an illness. While the stories at the beginning of this chapter illustrate how racial profiling can lead to deadly consequences, proponents of racial profiling in medicine do so in order to "diagnose disease more efficiently and prescribe medications more effectively"—that is, to save lives.[23] But despite the good intentions of clinicians and researchers, there are serious risks associated with the use of an essentialist logic in the medical arena, whether it is behind clinical misdiagnosis or superficial representation in biobanks.[24]

For example, like many people with mixed ancestry in the United States, I self-identify as African American, in part because of the historic construction of blackness as anyone with "one drop" of African ancestry, and because of my cultural upbringing in a society in which I was routinely identified by others as black. If I donated tissue for a stem cell line so that African Americans would have a better chance of finding an HLA match in the future, my maternal Iranian ancestry would likely undermine the prospect of African Americans with different ancestry to find a match and perhaps hide from Iranian Americans the possibility that they might find "my" stem cell line a resource.

Even so, it is vital to emphasize that analysts like myself who question the efficacy of medical racial profiling do not seek to whitewash the pressing racial health disparities that are an inevitable outcome of social stratification, or to pretend that the lived reality of race does not correspond to very serious life-and-death outcomes. Rather, we argue, the use of race as a proxy for genetic differences provides a simplistic picture of the connection between race and health disparities, and glosses over the social and environmental determinants of health as well as the way in which U.S. Census categories all too often hide the complexity of ancestry. Even the Hastings report notes that "within a family there is variability in HLA expression," so that while diversifying stem cell banks may increase the chances of someone who is not white finding a tissue match, there is no guarantee that those who classify themselves within the same census category will share the same ancestry, much less the same genetic makeup. This is even more true for African Americans, because "persons of sub-Saharan African ancestry have a greater variety of HLA types than do persons of any other geographical or ethnic grouping," thus requiring that much more genetic diversity to increase the chances of a match—diversity that does not necessarily correspond to sociopolitical definitions of race.

What's more, even as researchers are concerned about immune rejection, the problem of researchers being rejected by "hard to reach" populations in the context of recruitment continues to pose a significant barrier. For them to effectively culture stem cell lines, in other words, they must first cultivate consent (if not trust) in the research process. But recruitment efforts based upon the imperative of fair "biological access" to future therapies present a paradox: a vast literature documents that "treatment of African Americans by medicine in the U.S., from the period of slavery and well into the 20th century, was predicated on a scientific view, which posited that they were significantly biologically different from the white people."[25] That is, biological essentialism was explicitly used in the past to subordinate and harm racialized groups, whereas now it is implicitly used as a basis for including them and addressing their health concerns.

Trust Through the Prism of Whiteness

Prior to the Diversity Workshop, CIRM commissioned a study entitled "Supporting Diversity in Research Participation: A Framework for Action," which the author, Emily Friedman, summarized at the workshop. Among the explanations she provided for difficulties in recruiting diverse donors and study populations, there were the usual suspects: people's lack of understanding about clinical trials, lack of outreach on the part of trial sponsors, literacy issues and logistical barriers, and finally, "a general lack of trust in the health care system and especially in clinical research." The report also found issues specific to certain groups; namely,

> African-Americans tend to have the lowest level of trust in the health care system because of historical abuses. Chinese-Americans also have trust issues, as well as problems with English and, for older members of the community and recent immigrants, a lack of understanding of the underlying concepts of clinical research. Latinos also face language barriers, as well as a fear on the part of immigrants—legal or otherwise—that participation could bring negative consequences for them and their families. Southeast Asians share many of these issues, along with, for many groups, a fear of authority bred by a variety of traumas.[26]

What is striking is that despite the fact that *all* major ethnic minority groups are listed as having "trust issues" (the chronic omission of Native Americans notwithstanding), the report, the workshop, and human subject recruitment in general insist on framing "trust issues" as the exception, as "specific to certain groups"—when in fact their own findings show that it is much more the rule than the exception, except perhaps for white Americans.

More importantly, embedded in the "hard to reach" framing of minority groups is the idea that their relative distance from biomedicine is a matter of self-selection, rather than of systemic dispossession. While references to African Americans' "historic distrust" (and the habitual mention of Tuskegee as the touchstone for distrust) is at least a nod to some external justification for why people are "hard to reach," the ongoing social marginality that African Americans and other groups experience remains out of view in such hindsight statements. In fact, the large-scale survey of trust mentioned above showed that "though the legacy of the Tuskegee Syphilis Study is frequently cited as a major factor underlying distrust among African Americans, it did not appear to be related to societal or interpersonal distrust in our analyses."[27] In fact, the constant reference to historic distrust can divert our attention from the ongoing social production of distrust in daily encounters with the healthcare system and beyond. To reiterate what University of Wisconsin biologist and legal scholar Pilar Ossorio urged at the Toward Fair Cures conference, we need to shift the focus away from minority distrust to the "trustworthiness of institutions"—in the present and not simply the past, I would add.

Why then is distrust routinely conceived of as an anomaly to be "overcome" rather than a perfectly rational, perhaps even incisive, disposition toward biomedicine, based on knowledge that grows out of people's experiences with a host of institutions in a racialized society? To understand the persistence of this institutionalized myopia, we must note how well-meaning researchers perceive what we might term "white trust" as the norm by which all other groups are measured. Perhaps if we examined "white trust" as the problem to be overcome, insofar as it tends to hide the many shortcomings of free-market medicine to which "black *dis*trust" seems at least somewhat attuned, we would be forced to take on a different, far more complex understanding that might move us beyond offering free child care and translators for potential recruits. Not surprisingly, the benevolent "whiteness" that infuses CIRM's research and outreach recycles the same list of interventions found in other corners of the biomedical recruitment field, without fully addressing the trustworthiness of institutions.

Even so, not all of the interventions proposed at the Diversity Workshop were cosmetic; some were in fact more substantive, insofar as some participants recognized the need to base trust on an ongoing set of relationships. While CIRM's report suggested the use of "familiar faces" like celebrities, community leaders, and community-based physicians to communicate with targeted groups, there was also a more laudable emphasis on "early outreach," whereby community partners would provide input in the development of the research *before recruitment even begins* and "ongoing interaction" to "ensure that researchers do not 'helicopter' in and out of projects." In short, researchers were urged to "build relationships" and "long-term partnerships" with community leaders and local physicians rather than simply rely on the slick packaging of celebrity spokesmen. Dr. Lyndee Knox described Los Angeles Net, a nonprofit organization involving twenty community health clinics and 165 practices with over one million patient visits per year, in which "clinicians want to be engaged as team members and not simply viewed as recruitment sites." She detailed "the opportunity costs" of recruitment with the following breakdown: "A provider has 15 minutes or less per patient visit; this figure translates to 32+ patients a day per provider; If 3 minutes per patient are added for research, then 96 minutes are not available for primary care; as a result, 6.4 patients may not be seen in a day."

Her point was that researchers who seek to collaborate with Los Angeles Net and other community clinic networks must be willing to invest in the necessary infrastructure so that clinicians can at least sustain their existing quality of primary care (which, we should note, is already extremely rushed).

Finally, participants at the Diversity Workshop gave serious thought to the likely risks and costs for participants who require "lots of tests and close follow-up" and who would need to undergo a very rigorous consent process. By the end of the workshop, it seemed that community-based clinicians, not the patients themselves, needed to be won over. Indeed, at least one study indicates that hardly any existing "culturally informed" recruitment techniques have been effective. Rather, physician referrals have proved to be the single most effective strategy for enrolling African Americans.[28] But as my experience at the Garvey Research Complex made clear, primary care clinicians are often at odds with researchers, surgeons, and specialists, sometimes harboring their own brand of distrust that then impacts their role as gatekeepers to research initiatives.

Clinical Gatekeepers

Tate Wright, whom I introduced briefly in Chapter 4, regularly voices his frustration with "patient noncompliance" vis-à-vis standard sickle cell treatments but then often follows that up by saying he understands

that it was due to either the stress of his patients' daily life or the in-
effectiveness of prescribed treatments. As a white doctor in a context
where most of his staff is white but most of his patients are black, he
is also sensitive to the racial asymmetry and how that could inhibit patient trust and compliance. One afternoon in the sickle cell clinic, for
example, he shared the story of fifteen-year-old Tyrone Hemmingway.
Tyrone's family had elected to remove his spleen because, as is common
with sickle cell patients, it could get swollen when sickled hemoglobin
blocks the blood vessels. Wright explained that

> [t]he doctor who performed the surgery decided to use some high
> tech equipment that would allow him to do a laser surgery, which
> meant that he wouldn't have to open Tyrone all the way up. But
> because Tyrone's spleen was so big and they couldn't finish the op-
> eration in a reasonable amount of time, they left him open, iced his
> stomach, and wrapped him up. They brought him back to complete
> the surgery the following day, because they didn't want to keep
> him under anesthetics so long in his condition. But in the second
> round they accidentally lacerated his stomach, although they didn't
> yet know it. So they sewed him up, and when he got back into
> his room, he was complaining of pain. Staff thought it was just
> the post-op pain. Then he started peeing black urine and vomiting
> up blood. So they rushed him back into the operating room and
> opened him up and found the tear in his stomach. A few days later,
> Tyrone was still saying he was in a lot of pain, and it turned out that
> although they sewed the tear, he was digesting his pancreas.[29]

Wright continued the story, recounting several more operations that
were required to rectify the initial surgical damage and noting that in
all Tyrone had half a dozen surgeries in three weeks. He said that when
he went over to see Tyrone, he "honestly didn't think this kid was going
to make it. But he did. He hasn't been able to eat food for like three
months, and he was going out of his mind up there in his room, but
he pulled through, which is a miracle." Then, perhaps suspecting my
follow-up question, he mused, "and the family has been *so good* about

it. I mean they are angry, but they're not enraged like they *ought* to be. I mean, it was an elective surgery, and the poor mom, every time she left the hospital they called her to say that Tyrone was being rushed into the emergency room."[30]

I learned that it was typical of Dr. Wright to interject comments as he was going over a patient's history, remarks that expressed his disdain for the careerism of his colleagues. In one field note excerpt entitled "Hemoglobin Garvey," I recall how he was in the middle of explaining the differences in hemoglobin types when he noted that a new hemoglobin had been discovered at the Garvey Research Complex:

> He [Wright] noted wryly that it was named Hemoglobin Garvey, "which is a little tacky since you're supposed to name it after the patient, *not* the place that discovers it." A new medical student who had just joined the clinic followed up by asking the doctor a dozen questions about the technology involved in determining types of hemoglobin. After patiently answering, the doctor said that "a lot of what's driving the discovery of hemoglobin types is scientists trying to get their report in *Blood* [a medical journal], even if it's just based on one patient and the type is never seen again."[31]

From these and other accounts, I sensed this doctor's reluctance to refer his patients to other specialists whose decisions he often questioned—much less to sign them up for experimental procedures like the stem cell transplant that was then being offered in the research wing of the hospital.

Charismatic Collaboration

If we place Tate Wright low on a trust continuum, then Richard Gaskin, an African American stem cell activist, is so high as to be almost off the charts. Gaskin's rap moniker, "Professir X," draws upon the X-Men comic book saga—about mutant characters with abilities that can be both powers and liabilities—to use his disability as a resource in what activist Don Reed calls the "stem cell battles." Gaskin is a resident of Montclair, New Jersey, and was paralyzed from a gunshot wound in New

York when he was twenty years old. Since then he has worked with Michael J. Fox, Ted Kennedy, and Dana Reeve (wife of the late Christopher Reeve) to generate awareness and funds for stem cell research. Christopher Reeve's activism is what kick-started Gaskin's own advocacy career: "Before, there was nobody famous who represented me, except maybe Teddy Pendergrass. . . . here [referring to Reeve] was somebody who was going out there, fighting for a cure, advocating for better quality of life for people with disabilities, something I'd seen no one else do."[32]

Following Christopher Reeve's death, Gaskin performed a song entitled "Forever Superman" which Rutgers professor Wise Young, founder of the W. M. Keck Center for Collaborative Neuroscience, heard, thus launching a long-term friendship and collaboration. Young has been known to invite Gaskin up to the microphone in the middle of public lectures to perform one of his stem cell anthems, thereby "bringing a hip-hop vibe to the world of SCI (spinal cord injury) education and advocacy."[33] Gaskin has traveled with Young to China to see up close some of Young's clinical trials that use umbilical stem cells and lithium—leading one commentator to observe that their relationship "is not one of healer and patient. They work side by side as activists" to institute clinical trials in the United States for the over one million Americans with spinal cord injuries and countless others who experience paralysis. In Young's words, "This was unacceptable. How far have we declined in this country that we have to send people to China to participate in clinical trials of therapies developed in the U.S.? It isn't that umbilical cord blood cells and lithium are at all controversial. The only obstacle is money."[34]

The reality that Americans must engage in medical tourism rather than have access to cutting-edge therapies at home motivates both Young and Gaskin's advocacy and fund-raising. "The cost of holding clinical trials—which includes admitting 240 people [which is the typical size of a phase 1 trial] into the hospital; tests and treatments; and months of physical therapy—will be about $32 million. So Young and others came up with the JustaDollarPlease.org campaign, asking families and friends of spinal cord injured to give a dollar a day ($365 a year) and everyone to give whatever they can."[35]

Through their unique collaboration, Young and Gaskin demonstrate one way that "community collaboration" can generate support for and trust in stem cell research. It draws upon the charisma of a well-known figure like Gaskin, who can serve as both a spokesman and an educator, demystifying the research and providing the moral impetus for cures. This form of outreach is prominent in many discussions about how best to recruit "hard to reach" populations for clinical trials. It depends on the charisma of a spokesman who can speak not only *to* but *for* the community—translating the goals of research into everyday healthcare concerns while also reassuring researchers that community members trust them to work in their best interest.

It is important to keep in mind, however, that the practice of speaking for others is not straightforward or without its hazards.[36] Political scientist Michael Saward's work on representative claims-making rightly conceives "representation as a creative process that spills beyond legislatures," so that representatives are not simply those who have been elected, but anyone who seeks to speak on behalf of others.[37] For sociologist of science Michel Callon, "to speak for others is to first silence those in whose name we speak," even as "the groups or populations in whose name spokesmen speak are elusive."[38] In this context, spokesmen like Gaskin do not simply represent, but help to *produce*, the community's trust in Young's stem cell research. This is the creative dimension of spokesmanship; Saward is referring to it when he asserts that the "central aspect of political representation—the active making of symbols or images of what is to be represented" is a more accurate point of departure than a "fixed, knowable set of interests" which spokesmen either do or do *not* adequately represent.[39] His observation suggests that there is an element of "ventriloquism" in charismatic collaborations,[40] wherein spokesmen both represent and fashion the social world—those bonds of trust and accountability—through symbols, language, and emotion.

In his discussion of the symbolic order of society, sociologist Orville Lee critiques conventional political theory, noting that it "has shown itself to be considerably less able to account for the constitutive force of symbolic order and its effects on the democratic organization of society."[41] This division of labor, in other words, wherein cultural theorists deal in discourse, meanings, and symbols while political theorists speak of institutions, policies, and everyday behavior, neglects the deep connection between the two. Charismatic stem cell spokesmen conjoin rhetoric and reality, the symbolic world and the material, effectively intertwining the pragmatic requirements of basic research (epitomized by Young) with the poetry needed for clinical recruitment (epitomized by Gaskin).

While Richard Gaskin does not share many of the racial and class privileges of the predominantly middle- and upper-class white stem cell advocates we have discussed in previous chapters, he has, from the beginning of his career as a stem cell advocate, managed to speak on behalf of others. Inspired by the heroic figure of Christopher Reeves, in his painstaking effort to cultivate widespread support for and trust in this promising field, he, like other stem cell spokesmen, has utilized language and symbols to creatively reframe a quest for cures from an elitist privilege into a heroic cause.[42]

Competency Versus Humility

So far we have seen how researchers have approached the challenge of culturing diverse stem cell lines by cultivating community trust through the use of clinical gatekeepers, charismatic collaboration, and other forms of outreach. I have argued that in this process it is vital that the research community depathologize "black distrust" and question the normalcy of "white trust." These implicitly racialized dispositions are institutionalized in contemporary biomedical and public health discourses about "patient noncompliance" and "scientific literacy," and are firmly entrenched in the deficit model of public understanding of science, wherein "hard to reach" populations are routinely depicted as uneducated, uninterested, or even hostile. In short, the pathologization of those who do not avail themselves of emergent biotechnologies and therapies constitutes the normative underside of innovation. These deeply entrenched associations, in turn, need to be tackled on the institutional, and not simply the interpersonal, level.

In contrast to the deficit model underlying much biomedical and scientific outreach, analysts have identified the multiple ways in which people engage with science as "competent human subjects" who have their own "civic epistemologies" from which they evaluate developments in and deployments of science and technology.[43] This prefigures the paradoxical situation I found during my fieldwork, wherein new facts about biological processes and new techniques for intervening in the body paradoxically produced new *ignorance*. In their confrontation with uncertainty, research participants must weigh, situate, resist, and integrate new facts *and* new ignorance in a vulnerable context.[44] Perhaps that is why medical anthropologist Rayna Rapp refers to those on the biomedical front lines as moral pioneers: "[a]t once conscripts to technoscientific regimes of quality control and normalization, and explorers of the ethical territory [biotechnology's] presence produces."[45] But for Rapp, these moral pioneers are not only those who avail themselves of biotechnologies but also those whom she refers to as "refusers" and "draft resisters."[46] Both recruits and resisters alike engage in a "philosophy of the limit,"[47] wherein they enact personal choices within a context of social limits beyond which they are unwilling or unable to tread.

As Rapp explains, these limits to choice are mediated by gender, ethnoracial, class, and religious structures—reminding us of Karl Marx's oft cited correction to the ideology of free choice: "People make their own history, but they do not make it as they please; they do not make it under self-selected circumstances, but under circumstances existing already, given and transmitted from the past."[48] People engage with biotechnologies through a complex interplay of their past experiences (real and imagined), present circumstances, and future hopes and fears—a dynamic introduced in the Introduction with the notion of "Sankofa science in action." As with each of the subsequent struggles to constitute the "we" of people's science—in this case either free-floating, trusting individuals who are optimistic about the future, *or* socially situated collectives looking ambivalently over their shoulder—the way we go about enacting public participation in science shifts depending on how we imagine "the public." In a society that presumes and cultivates autonomy and free choice, those who resist modes of biological citizenship as enshrined in the stem cell initiative are vulnerable to the charge of defection: in refusing to accept their responsibility as research recruits, they also fail to help rid the body politic of biological defects. It is not enough then to simply acknowledge that "refusers" are justified in their distrust of the medical establishment. After all, the practice of acknowledging distrust in a sympathetic vein has become commonplace among clinicians and researchers. A substantive approach to the issue requires a complete reorientation to it—away from the ethnoracial and cultural identities of "problem people," to borrow W.E.B. Du Bois's phrase.[49] Rather, we must challenge the "cultural competency" model of biomedical inclusion in which medical practitioners are often taught to master a finite body of knowledge (laundry lists of cultural traits, for example) in order to effectively relate to their patients. Oakland, California–based physician Melanie Tervalon argues that "cultural humility," not competency, is a more suitable framework: "Cultural humility incorporates a lifelong commitment to self-evaluation and self-critique, to redressing the power imbalances in the patient-physician dynamic, and to developing mutually beneficial and nonpaternalistic clinical and advocacy partnerships with communities on behalf of individuals and defined populations."[50] Tervalon rejects the deficit model of expert-lay relationships insofar as it demeans the competency of so-called laypeople, while ignoring the subjectivity and biases of experts. Too often, biomedical culture and white American values of autonomy and free choice elude interrogation, while the cultures of "diverse" patient populations are regarded as exotic or problematic.[51]

It is one thing for individual clinicians to cultivate self-awareness as a prerequisite to practicing medicine; but I am suggesting that institutions, too, such as California's stem cell agency, grow more self-reflective about the normative underpinnings of their aims and practices. The structures of participation, modes of inclusion, and assumptions of advocates require explicit identification and thoughtful evaluation with respect to their shortcomings as much as to their promise. Developing *technologies of humility*—that is, "disciplined methods to accommodate the partiality of scientific knowledge and to act under irredeemable uncertainty"—redirects our attention to the ethical and normative dimensions of scientific knowledge and governance.[52] Such technologies compel us to begin redressing inequality before setting out to produce (which often means in practice simply *re*producing) information about why subordinate groups remain elusive to researchers. If we understand trust and distrust not simply as individual or cultural predispositions that are "held" by some and not by others but rather as outgrowths of social relationships that are produced through the allocation of material resources and symbolic power, then we see that techniques for cultivating trust hinge on redistributing and refashioning those, respectively.

CHAPTER SIX

TOWARD REAL UTOPIAS

There is a sense in which all technical activity contains an inherent
tendency toward forgetfulness.

—*Langdon Winner*, Autonomous Technology[1]

What good is having the right to sit at a lunch counter if you can't afford
to buy a hamburger?

—*Martin Luther King, Jr.*[2]

Let's say the research goes on. The drugs are developed. They're very
expensive, and what good has it done you if you can't access them?

—*Senator Sheila Kuehl*[3]

ON DEC 1, 2004, the television host and political commentator Rachel
Maddow hosted a radio program, *Unfiltered*, in which she dedicated the
regular segment called "Burying the Lead" to a California ballot measure that she suspected very few people knew about. In this case, Maddow was not referring to California's Proposition 71 but Proposition 72,
a bill that would require large- and medium-size companies to provide
health insurance to their employees and pay at least 80 percent of the
cost. At the time of Maddow's reporting, the ballots were still being
counted, but the voting seemed to be turning in favor of the Yes on 72
campaign, which ultimately came up short, with 49.2 percent yes votes.

Perhaps guilty of burying the lead as well, I have waited until nearly
the end of this book to fully question what a "right to stem cell research"
means in a context in which basic healthcare access is not yet a guar-
anteed right in the United States. For whose bodies have we codified a

right to develop novel cures? Presumably, the de facto constituency of Prop. 71 are those people who already have access to existing treatments, because for those who do not, a right to basic healthcare would necessarily come before a futuristic investment like stem cell research. This tension between elite and populist priorities led one observer to frame the larger implications of Prop. 72 as "the most important sleeper on the ballot this year":

> Without knowing much about it, people don't realize they may be voting on the hope of national health care reform. Prop. 72 [may] sound boring, but [it's] really one of the most important battles in the country this November. It's David vs. Goliath on health care reform. . . . Why should you care? Because huge companies see it as a litmus test for national health care reform, and they want to stop it. . . . Can you imagine a world where people weren't chained to jobs because of insurance? Can you imagine actually turning back the tide of healthcare inflation? Can you imagine standing up to giant corporation machines like Wal-Mart and finally winning? That's what's at stake, not just here but for the country, on Prop 72.[4]

As for those "giant corporations," top donors to the No on 72 campaign included the California Restaurant Association, Wal-Mart, and McDonald's, as well as a number of other national chain stores and restaurants,[5] while top donors to the Yes on 72 campaign included the Service Employees International Union, the California Teachers Union, the Food and Commercial Workers Union, and the California Healthcare Association.[6] This showdown reflects what pundits worry is the rising "class warfare" in the United States, as if the policies that supported the accumulation of wealth in the hands of a smaller and smaller subset of the population over the last several decades were not in themselves an assault upon the opportunities and life chances of the majority of Americans.

Exemplifying the age-old struggle between the strong and the weak, a biotech firm actually named Goliath worked to mobilize opposition against Prop. 72, threatening that "in the end, the very people that

[Prop. 72] was attempting to help—employees without health care—will not only *not* have health insurance, they won't have jobs."[7] Another No on 72 advocate offered the following scenario, repeated in various forms in the days leading up to the November 2 election:

> Consider a minimum-wage employee with minimal health insurance coverage. Since his salary cannot be cut any further by law, terminating him would be the only way for a business to cope with the costs of this mandate. As a direct result of Proposition 72, this employee would not only fail to see any improvement in his health coverage, he would lose his job and the minimal coverage he had previously enjoyed. Likewise, a business with 25 employees would be faced with a choice of shouldering the unsustainable burden of Proposition 72 and facing bankruptcy or laying off 6 of its employees to become exempt. It doesn't take a CEO to realize the route this hypothetical business will take.[8]

Other biotech companies followed suit, joining in a campaign against the "economic catastrophe" of Prop. 72 while throwing all of their support behind Prop. 71.[9] One after another, they expressed concern about the country's healthcare crisis but insisted that Prop. 72 was *not* the answer. Bill Goodrich, president and CEO of the United Agribusiness League (UAL), the voice of the agricultural industry, argued that "[a]n answer to the uninsured population clearly needs to be found. But with California's current economic difficulties, now is not the time to burden our businesses with employer-mandated healthcare coverage."[10] The public was warned of the pitfalls of "government-run healthcare," despite the fact that numerous labor, medical, and consumer groups tried to counter that false charge.[11] The California Black Women's Health Project, among others, clarified that "Proposition 72 is not a government-run healthcare system. It is private insurance, paid for by employers for employees. Prop. 72 provides a state purchasing pool for employers who choose not to purchase health insurance directly. Over one million uninsured Californians will stop depending on taxpayer-paid care for the uninsured and start getting private insurance paid for by employers."[12]

While I am not convinced that Prop. 72 would have had a sustainable, long-term impact for those who most need access to quality healthcare or that it could avoid the threat of increasing unemployment due to employers' cutting costs, its defeat alongside the passage of Prop. 71 does reflect an underlying tension between elite and populist "rights": investment in future medical breakthroughs versus expanding access to present public goods. In light of other public spending priorities, one observer explained it this way:

> While $300 million per year [the budget of the stem cell agency] is only one third of one percent (0.35 percent) of the State's $86 billion General Fund budget, it is nonetheless a misplaced commitment when State Medi-Cal and Workmen's Compensation funds are being drastically cut. For example, the State Legislative Analyst has recommended that funding for visits to physicians and treatment centers could save $196.5 million per year by limiting visits to 10 per year. . . . But the self-perpetuating California Center for Regenerative Medicine is standing in the way. We can no longer afford luxury jobs programs for biomedical professionals for hypothetical research which is already amply funded by both the private sector and the National Institutes of Health, while medically needy people are in need of resources for care in California.[13]

But while many Prop. 72 opponents expressed concern for "medically needy people" and some even granted that the state has a responsibility to ensure that they receive healthcare, ideas for how to achieve this were scarce. Instead, proposals to build in accountability for equitable healthcare access, specifically in the context of the stem cell initiative, were repeatedly shot down by most members of the Independent Citizens' Oversight Committee and the agency's most loyal supporters.

For example, in the year following the passage of Prop. 71, California senator Deborah Ortiz was at the forefront of sponsoring legislation to regulate the stem cell agency.[14] Stem cell advocates routinely criticized these legislative efforts as power-hungry meddling, as when physicist and Nobel Prize winner Paul Berg commented that "this is a whole new

structure laid on top of the state government, which has the responsibility for giving away three billion dollars. And as you might imagine, the
legislators are very, very envious and jealous."[15]

The most controversial bill was SB 401, which sought to tighten
regulation around CIRM's intellectual property policy. CIRM opposed
the bill, in part because "language directed towards increasing the share
of revenues to be allocated to the state from Prop. 71's future successes
may have the effect of discouraging applications for research grants,"
and because "participation by important research groups and entities
could be [negatively] impacted by tying what they can charge for their
costs, should a product (drugs, therapies, etc) be developed, to federal
Medicaid pricing levels."[16] Reacting to the proposed changes, Americans
for Cures circulated a letter to supporters that called Deborah Ortiz "an
ongoing threat to Proposition 71," explaining that

> [i]n a stream of legislative acts, Senate Bill 18, Senate Constitutional
> Amendment 13 and most recently Senate Bill 401, the Senator has
> attempted to impose crippling restrictions on the research she
> claims to champion. Rather than allowing the CIRM time to de-
> velop policy and standards, Senator Ortiz attempted to impose her
> own legislative controls on the new agency barely one month after
> its approval by the voters. Her policies, if implemented, would be
> disastrous for the new stem cell research program. SB 401 will stunt
> progress in stem cell research. . . . Above all, SB 401 ignores the
> single most important element in stem cell research: the patients
> and their families.[17]

Senator Ortiz also seemed to have patients and their families in
mind—specifically those who would find themselves unable to afford
stem cell treatments if expedience outweighed equity in the implementation of the initiative. At a June 6, 2005, meeting of the Independent
Citizens' Oversight Committee,[18] Ortiz explained that

> [w]e have a huge challenge every year to serve the poor in our
> healthcare programs, in our constantly shrinking budgets, and that

has been sort of the promise we presented to the voters in Prop. 71. It's that policy I think we remain in disagreement on how best to effectuate. . . . And then we balance that objective and that value against the need to assure that essential medical research is not unreasonably hindered by the intellectual property agreements. . . . What I think is the highest and most important objective in my efforts is to assure a stream of treatments to our poor in California.[19]

For many board members and stem cell advocates, the senator's demands were seen as premature, if not also overly restrictive and burdensome. Robert Klein requested the legislature to back off and give the stem cell agency the "ability to innovate and try and figure out how we can make these programs affordable."[20] Don Reed seconded this, suggesting that "greatness needs time to grow."[21] A few weeks earlier, without Ortiz present, Reed had been even more insistent:

We can be polite as we want; but if we're not clear and blunt and outspoken, she'll win, because the people in there, first of all, they like power. They like oversight. They want control. It's in their nature to control. And if she, the expert, whom they [other members of the legislature] know was supportive [of stem cell research], if she says control is needed, their first instinct is going to be to go with her. . . . The attempt to make this affordable has been tried before. The NIH has tried it, and there's a good study, which shows that it was an utter disaster. What people came to realize is that before we can have affordable computers, we must have computers. The greater good is the benefit of this, not the small individual tinkering, which will slow the whole thing down.[22]

Another member of the public, Denise Reynolds, whose good friend was paralyzed in a biking accident, implored, "Let us not become greedy in demanding greater revenue and outcost treatments to the low-income populations."[23] ICOC member and Burnham Institute director John Reed added that "[t]o me (SB 401) is death by 1,000 cuts."[24] Similarly, David Baltimore, president of the California Institute of Technology and a board member and investor in several well-known biotech companies, addressed Ortiz directly at the meeting:

> You are trying to burden this bill with a huge social problem in America, which is the disparity between the healthcare available to the poor and the healthcare available to the rich. And I am very sensitive to that and totally supportive of trying to do something about it. But if you burden an initiative whose focus is research with solving that problem, then you get yourself involved in all of these very detailed issues, and at the same time it is a snare and a delusion to believe that the resources that are going to come from this research are going to be able to pay for the needs of the poor. . . . We're talking about something that's simply not going to exist. You're worried about the return to Californians. Research is not about financial return. Research is about setting the basis for therapy. . . . I would wish that somebody with your obvious sympathy with the need for research would cheer on this group rather than tying us up so that we are unable to carry out the function that was provided for us through Proposition 71.[25]

Perhaps the idea that "research is not about financial return" was a truism forty years ago, but in the wake of the Bayh-Dole Act and the privatization of research, Baltimore's earnest appeal rings hollow. The question is not whether the research should seek to be profitable but rather *profitable for whom?* The agency was doing all it could to ensure that private biotech companies would profit from stem cell research, because they were seen as a crucial gatekeeper in the race for cures, even as the act itself recognized "the conflicting interests involved in IP [intellectual property] by providing that 'The ICOC shall establish standards that require that all grants and loan awards be subject to intellectual property agreements that balance the opportunity of the State of California to benefit from the patents, royalties, and licenses that result from basic research, therapy development, and clinical trials with the need to assure that essential medical research is not unreasonably hindered by the intellectual property agreements.'"[26]

When SB 401 was eventually rejected and Senator Deborah Ortiz had reached her term limit, Senator Sheila Kuehl replaced Ortiz as a primary antagonist of the agency, going on record as saying that "[b]asically, the attitude has been that they just want the Legislature to go away so they can run their own show."[27] She warned that the agency's oversight proposal was "a weak and vague standard that [was] unlikely to result in any meaningful access for the uninsured to new stem cell drugs and therapies. The proposed regulations actually take a step backwards from previous iterations, by allowing the [biotech company–sponsored] access plans" to police themselves.[28] To address this, Kuehl introduced SB 1565, a proposal that would require the agency's intellectual property standards to

> include a requirement that each grantee and the licensees of the grantee submit to the CIRM for approval a plan that will afford uninsured Californians access to any drug that is, in whole or in part, the result of research funded by the CIRM, and would require that any plan subject to that approval shall require that the grantees and licensees thereof sell drugs at a price that does not exceed any benchmark price in the California Discount Prescription Drug Program.[29]

Although it passed by a vote of 16–0, Governor Schwarzenegger eventually vetoed the bill.

Reacting to ongoing antagonism between the stem cell agency and the California legislature, consumer watchdog advocate John Simpson complained that "[p]art of the problem with this whole thing has been an ongoing attitude of, 'You're with us or you're against us and you're against stem-cell research.' That's just nonsense. It ought to be the case that you can have constructive criticism of things that a state agency does without people shouting back at you that you're against stem-cell research."[30] "When they get responsible criticism from outsiders, they kind of circle the wagons and say, 'We know science, we know best.' I think they would be much better off if they would sit down and engage."[31] In many ways, the circle-the-wagons solidarity that had helped mobilize a disparate group of patient advocates, scientists, Hollywood celebrities, and biotech investors in the passage of Prop. 71 had become a liability. The zealous pro-cures and proscience campaign rhetoric that had made for effective sound-bite politics was now alienating potential allies at a time when more tempered and inclusive deliberation was required to implement the initiative.

We know that unlike his predecessor, Barack Obama has been a major supporter of stem cell research, using his first act as president to loosen federal regulations on the struggling field. Despite the red-blue polarization for which SCR is known, it is Obama's national healthcare bill that has garnered the fiercest criticism and backlash from political opponents. How then do the bioconstitutional struggles over equity and inclusion in stem cell research fit into the broader U.S. healthcare debate?

It is a challenge to address this, in part because the current U.S. health system is undergoing shifts that in many ways mirror Prop. 72's effort to mandate health insurance via employers. The Obama administration has created an imperative for individuals and families to purchase insurance, with a significant subsidy for those who make between 100 and 400 percent of the federal poverty level, alongside other major provisions. Reflecting on the impact of the current administration's health reform, *New York Times* columnist David Leonhardt reported that

> For all the political and economic uncertainties about health reform, at least one thing seems clear: The bill that President Obama signed on Tuesday is the federal government's biggest attack on economic inequality since inequality began rising more than three decades ago. . . . Much about health reform remains unknown. . . . Maybe the bill's attempts to hold down the recent growth of medical costs will prove a big success, or maybe the results will be modest and inadequate. But the ways in which the bill attacks the inequality of the Reagan era—whether you love them or hate them—will probably be around for a long time.[32]

Leonhardt goes on to frame the larger tension animating proponents and opponents of Obama's health reform as arriving at fundamentally different answers to "the central question that both the Reagan and Obama administrations have tried to answer—what is the proper balance between the market and the government?" While the 2010 health reform certainly shifts the balance toward the latter, its impact is as yet undetermined.

But even as we await the effects of federal health reform, the relationship between healthcare access, social inequality, and medical innovation is more apparent and can guide our thinking about public spending. An article on stem cell research and social justice pointed out that (prior to the impending reforms)

> as many as 46.6 million people in the United States lack health insurance. . . . Those without it are less likely to receive preventative care and thus to be in advanced stages of a disease once they are examined by a physician. They also tend to be sicker upon being admitted to a hospital and thus more likely to die after being admitted. In addition, even though those without health insurance tend to be poorer, they often have greater out of pocket expenses and have higher rates of bankruptcy due primarily to their medical expenses.[33]

Ethnoracial minorities, in turn, "are much more likely to be without health insurance than white Americans. In 2005, 11.3 percent of non-Hispanic whites lacked health insurance, while 19.6 percent of African Americans and 32.7 percent of Hispanics did not have it. The impact of not having health insurance on people's health and financial viability cannot be overemphasized."[34]

An analysis of National Health Statistics data found that between 1991 and 2000, almost 900,000 African American deaths could have been prevented if blacks had received the same healthcare as whites.[35] The study estimates that "technological improvements in medicine—including better drugs, devices, and procedures—averted only 176,633 deaths during the same period."[36] The authors thus questioned "the prudence of investing billions in the development of new drugs and technologies while investing only a fraction of that amount in the correction of disparities," given that the deaths averted through improvements in technology constituted only a fraction of those that could had been saved through access to basic healthcare. The lead author of the study, Steven H. Woolf, explained that "five times as many lives can be saved by correcting disparities than in developing new treatments."[37] A second author, Otis Brawley, explained that "it's important [to note] that this is not an argument against science. . . . This is an argument that there are therapies out there that are not new that people just don't get."[38] A similar study using 2002 data found that "an estimated 83,570 excess [African American] deaths each year could be prevented in the United States if this black-white mortality gap could be eliminated."[39]

From Center to Margin

Political philosophers and progressive educators alike have long emphasized the importance of storytelling in opening up the civic sphere to a broader range of publics. Rational decision-making, whether about health, safety, education, or fiscal responsibility, requires that the stories of otherwise marginalized groups enter and influence public discourse. Genuine empathy is a necessary feature of this discourse; impartiality is not about discarding feelings and vetting opposing claims through "dry logic." Rather, a robust civic sphere requires us to "take the position of the concrete other in order to judge problems *more* impartially."[40] The universal citizen—that abstract figure who exercises rights and bears responsibilities in the modern polity—is a myth. By this I do not mean that it is a baseless delusion, but that it is a story that we tell ourselves: an ideal that we desire but a fiction nonetheless that hides the many ways in which our current social order does not simply produce "outsiders within" society, but relies upon them.[41]

While important work in the last several decades has sought to bring marginalized perspectives into the center of public discourse and intellectual thought, the inverse process of considering major developments in the life sciences *from the periphery* is perhaps an equally vital shift. The public sponsorship of stem cell research in California is part of a longer and larger story about citizenship, about how priorities are established, about how economic value and collective values intertwine, and about how the future is imagined. What we have observed in the previous chapters is how, much too often, the few speak for the many: the elite speak for the masses, projecting the desire for future treatment onto people who cannot obtain *existing* therapies which could prolong and save their lives *right now*.

Citizenship, after all, is not simply a political status so much as it is a *process*.[42] The challenge for us is to enact health and science policy in a way that acknowledges what we share and have in common while at the same time taking account of important, often life-and-death social cleavages. Rather than opening up space within which to deal critically with the differential impact of health policy and priorities on different populations, public engagement with science is typically manufactured to recognize only a certain set of prescribed differences (such as those recognized in the ten disease advocacy representatives on the ICOC) that do not challenge the underlying prioritization of innovation over equity.

Beyond Bioconstitutional Struggles?

Finally, it is not enough to simply broaden the parameters of social struggle, allowing for a more "fair" contest between the haves and have-nots over biomedical goods (for example, the right to "fair access"), without at once querying the underlying assumptions of the struggle. Shifting attention from consumption of science and technology as the primary means of exercising citizenship, to participation as a more robust form of citizenship, in which people actively engage in the production of knowledge, forces us to contend with challenges that are even more entrenched than administering universal healthcare. In *Re-thinking Science: Knowledge and the Public in an Age of Uncertainty* (2001), Nowotny and colleagues explain that "in the eyes of many citizens, science and technology are now equated with their results and products. They have come to be seen largely as commodities, access to which should be democratically regulated and the allotment of which should also be fairly distributed. Consequently, what should be produced and how it should be produced must be embraced within democratic decision making."[43]

We have inherited "a social structure that has accepted separation as the norm—the dividing of people into groups of 'we' and 'they' who fight, who compete, who negotiate, who cooperate, or who help each other from across the boundaries that define their separateness."[44] I borrow this last insight from a prescient compilation of papers called *The Lab, the Temple, and the Market*, in which an interdisciplinary group of analysts reflect upon the role of science, religion, and economic development in human emancipation and social subordination. They question the role of scholars in reifying social differences and group struggle, and ask whether

> human beings [are] doomed to be outsiders to every group except a single one, a subculture narrowly defined by nationality, ethnicity, social class, religion, and occupation. . . . This tendency reinforces, and is reinforced by, an intellectuality that sees as the hallmark of intelligence the ability to identify differences, to divide, and to relativize, all in the name of being scientific. Such an approach is a gross misrepresentation of science, for although it is true that science analyzes, it also integrates and points to underlying patterns of oneness.[45]

In particular, the work of Farzam Arbab, physicist, educator, and founder, in 1974, of the Colombia-based Fundación para la Aplicación y Enseñanza de las Ciencias (Foundation for the Application and Teaching of the Sciences, or FUNDAEC),[46] informs my thinking about how we can imagine and enact more egalitarian forms of participation in science in a way that is liberating and just. While the Colombian and Californian contexts are dramatically different on a number of fronts, they are similarly ordered by the "current arrangements that assign the ownership of modern science to small sectors of society." If alternatives to this social order are not created, we can expect a "widening of the gap between the poor and the rich," which is as well a deeply racialized cleavage.[47]

In California, for example, where the poverty rate is already typically higher than the rest of the nation, Latinos (22.8%) and African Americans (22.1%) have a substantially higher rate of poverty than whites (9.5%) and Asians (11.8%).[48] The life-and-death implications of racialized economic inequality in turn are nowhere more evident than in mortality rates. In California, a black man "can expect to live 68.6 years on average, which is far below the life expectancy of the average California white male, who can expect to live 75.5 years,"[49] leading a Harvard University professor of public health, Nancy Krieger, to assert in no uncertain terms that "social inequality kills."[50] Her work, among that of others, demonstrates that popular mandates and movements for social justice (for example, the Civil Rights Movement and the War on Poverty, including Medicare and Medicaid, which began closing the racialized gap between rich and poor) were the primary drivers for alleviating health inequities between 1966 and 1980.

> Starting thereafter, in 1980, with the Reagan revolution we've had 25 years of conservative policies that have sought to roll back the "Welfare State," [and] rein in big government except, of course, for military spending. And we have seen this play out in a familiar litany of deregulating industry, slashing [taxes] on the wealthy, repeatedly freezing the minimum wage, and undermining funding for public health infrastructure. . . . The net result has been widening health inequities. *So are these inequities inevitable?* The clear answer is "no." Our data show that the inequities shrank and then they have widened. *What are the alternatives?* We as a nation have to say that we have the political and social priorities, because we *do have* the economic resources to address these inequities.[51]

In sum, a large part of moving forward entails remembering—Sankofa-like—what has worked in the past rather than allowing ourselves to be hypnotized by the latest shiny innovation promising to revolutionize health. But rather than bracing ourselves for a roller coaster of progress and retrenchment corresponding to changes in political administrations, we need to tackle a much more fundamental question: not simply, what kinds of policies and initiatives do we wish to undertake, but also, who are the agents of change?

Here I draw upon a new wave of sociological thinking about "real utopias," which challenges us to imagine and create institutions and social relations that are fundamentally different from the status quo.[52] As 2012 president of the American Sociological Association Erik Olin Wright describes it, a real utopia embraces the

> tension between dreams and practice. It is founded on the belief that what is pragmatically possible is not fixed independently of our imaginations, but is itself shaped by our visions. Self-fulfilling prophecies are powerful forces in history. . . . Nurturing clear-sighted understandings of what it would take to create social institutions free of oppression is part of creating a political will for radical social changes to reduce oppression. A vital belief in a utopian ideal may be necessary to motivate people to leave on the journey from the status quo in the first place, even though the likely actual destination may fall short of the utopian ideal.

He emphasizes that *real* utopias are "ideals that are grounded in the real potentials of humanity, utopian destinations that have accessible way-stations, utopian designs of institutions that can inform our practical tasks of muddling through in a world of imperfect conditions for social change."[53]

In a similar vein, at the Science and Democracy lecture series at Harvard in 2010, writer and activist Arundhati Roy exhorted participants: "The first step towards reimagining a world gone terribly wrong would be to stop making war on those who have a different imagination, an imagination outside of capitalism and, for that matter, communism, an imagination which has an altogether different understanding of what constitutes happiness and fulfillment."[54] Consider that throughout the previous pages you have been introduced to many people willing to invest money and time to hone scientific techniques that might defy (what were once accepted as) laws or limits of biology as they relate to aging and illness—as when journalist Teri Somers asks us to "[i]magine cardiac cells, beating in a petri dish, being used to form human tissue that might be used to replace damaged heart muscle."[55] But somehow imaginations go limp when we are confronted with social dis-order. For many people, the idea that we can defy politics as usual and channel human ingenuity toward more cooperative and inclusive forms of social organization is utterly far-fetched. Thus I am convinced that any institutional retooling must begin by first querying this faith in biological regeneration along-side our underdeveloped belief in social transformation. *If our bodies can regenerate, why do we perceive our body politic as so utterly fixed?*

That being said, I would like to spend the last pages of this book sketching a few principles for participatory science-making that, while not suggesting a fully fledged real-utopian blueprint, seek to outline several tenets that I consider crucial to any such effort. The proviso is that it would be antithetical to any such institution-building to offer directives; rather, I offer these principles as raw material for designing "technologies of humility"[56] that must ultimately be the subject of wide-ranging consultation as we create alternatives to top-down, elite-driven science and technology policy. Like Farzam Arbab, I am concerned about "the right of the masses of humanity not only to have access to information [or therapies], but to participate fully in the generation and application of knowledge."[57] To that end, I have distilled five lines of praxis from the otherwise far-reaching essays in *The Lab, the Temple, and the Market*—practices which I think are essential pillars in the construction of real-utopian designs for participatory science.

Not Equating Winning with Being Right. Democracy as it is currently imagined and enacted divides people "according to interest, talent, and ideology, who then 'negotiate' decisions," wherein "the purpose of each component group is to win. The means to this end are economic advantage and the mobilization of support to overwhelm the opponent. So strong is this legacy of 'he who wins is right' that it essentially determines the way justice is administered."[58] We see these dynamics at play in the Prop. 71 and 72 campaigns, as well as in the implementation of Prop. 71, wherein those who monopolize capital were best positioned to materialize their particular vision of the good life. How often did the proponents of stem cell research use the ballot win to assert that "the people of California" gave them a mandate to pursue expedience over equity? Their "win" in turn bolstered their opposition to incorporating forms of equitable distribution of stem cell goods. Arbab urges us to question whether we should accept this idea that "he who wins is right" as the "crowning achievement of the evolution of collective decision-making on the planet."[59] To which I answer, emphatically and optimistically, *no*. The fierce competition we take for granted in "democratic" life, wherein those who monopolize capital wield inordinate power and influence in the name of the many, is an outmoded form of top-down governing. Even when the winner *doesn't* take all, "losers" are still left to squabble over crumbs, as when only 2–5 percent of Big Science initiative budgets is routinely allocated to investigate ethical, legal, and social issues related to the field. When public participation is staged, it often takes the form of the shallowest form of exchange, as when state agencies routinely stage "public hearings" at which speakers are allotted no more than three minutes to express their views, while final decisions are still monopolized by a select few, who often make them behind closed doors.

Cultivating the Art of Listening over That of Manipulation. Whether governing conventional spheres of life, such as housing or education, or more novel spheres, such as regenerative medicine, we can no longer rely upon the art of political manipulation as the taken-for-granted backdrop of decision-making. Rather, *a posture of learning* must inform and precede our collective investigations and rational analysis of options.

> Such an attitude is basically different from that of experts or highly paid consultants who generally act on the basis of a series of certainties coming from their "knowledge" or "professional experience." Such "authorities," particularly when they refuse to question their certainties, not only tend to mislead the people in whose lives they intervene, but also lose touch with the very objects of their knowledge. Because they are unable to listen, they find that their accumulated knowledge soon becomes obsolete and of little relevance to the changing realities they address.[60]

An example of public policy decision-making based upon thoughtful listening rather than manipulation can be found in the Interactivity Foundation's "citizen discussions."[61] As part of my fieldwork on the California stem cell initiative in 2007, I took part in a series on "Anticipating Human Genetic Technology," which started off with the expectation that those gathered were "an informed citizenry *offering impressions*, rather than a persuaded citizenry *offering opinions*."[62] A facilitator introduced a number of alternatives for thinking about the development of genetic technologies, and the group proceeded to unpack the implications and unintended consequences of pursuing each alternative; for example, we explored the concerns around and benefits of "limiting human genetic technologies," "embracing human technologies," and "balancing social and individual control."[63] We then came up with suggestions for redesigning these alternatives, followed by broader observations and questions about each one. This process of consultation was neither wholly detached nor impassioned, but it deepened participants' awareness of the many sides of a policy issue (even if we may have initially come in with a strong point of view) rather than polarizing us in the way that electoral politics do.

The posture of learning that participants assumed as we collectively explored the specifics of various institutional designs was cultivated in part through the Interactivity Foundation's deliberation tools, which were expertly applied by the facilitator. That is to say, simply putting "diverse" people in a room with the "freedom" to express their viewpoints is not an effective or sustainable alternative to top-down policy decision-making. Mere inclusion of different points of view is not a panacea for the conflict-ridden politics to which many of us are accustomed; indeed, precisely because we are accustomed to it, we are likely to fall into habitual scripts in which vying for the upper hand seems "natural." Rather, the Interactivity Foundation asks prospective discussants to help one another "think through . . . ideas," not as advocates or competitors but by "step[ping] outside their own thinking and think[ing] about issues as people from different backgrounds might see them."[64]

Developing Consultative, Not Consumptive, Capacity. In thinking through the challenge of cultivating a wider capacity for decision-making beyond the privileged realm of legislatures, consultants, and think tanks, Farzam Arbab, in *The Lab, the Temple, and the Market*, explains that

> [i]n a world all too given to twisting words to suit economic interests, the capacity to make proper technological choices could easily become synonymous with the possession of the skills of a good consumer. This is clearly not what is intended here. The type of capacity under discussion represents a complex set of attitudes, convictions, understandings, skills, and habits, all of which characterize the behavior of individuals and organizations in their daily interaction with technology.[65]

The more that people are conceived of and treated primarily as consumers and end users, the less their creative capacity to imagine and produce alternatives can be realized.[66]

The tension between consuming and producing the fruits of science appears to also animate UC Berkeley professor David Winickoff's argument that we must seek not only equitable *benefit-sharing* but also equitable *power-sharing* structures: "While benefit sharing should be applauded insofar as it attempts to submit relations of biocapital to new claims of distributive justice, the project is likely to fail both as a normative and practical matter without greater attention to issues of procedural justice."[67] Again, the Interactivity Foundation offers one model for expanding decision-making and power-sharing capacity, in which we engaged in five stages of "citizen policy discussions":

1. Describe the area of concern by developing questions
2. Generate policy possibilities that respond to those questions
3. Explore possible consequences in order to revise the policy options
4. Organize joint panel discussions between persons with professional experience relevant to the area of concern and citizens whose familiarity with the area of concern grows out of their general life experi-

ences outside of work (typically called "lay" persons, their experiential knowledge is no less vital to the process)

5. Create a citizen discussion report (and, I would add, items for collective action)

This is not a process that can be rushed through in one session or during a single conference, however well intentioned—and certainly not in the kinds of public hearings that are the current hallmark of "open" and "transparent" governance at California's stem cell agency. When we consider the often slow, deliberate, incremental process by which scientific knowledge is typically generated, why is it that we do not expect or implement the same kind of attentive dynamics in *social* experiments that seek to generate new knowledge? Rather, we have a deeply asymmetrical approach, in which our investment of both time and money in reengineering biological life far exceeds our collective will to transform *social* life.

Fostering Reciprocity Between Institutions and Individuals. It is vital that institutions cease imposing the views of a particular faction or class, whether democratically elected or not, on the rest of society. Conservation biologist and MacArthur fellow John Terborgh has argued that "[t]he public, and especially the political class, cherry-picks its science. If a scientist finds a promising treatment for AIDS or cancer, then he is a hero; if he warns about overpopulation, climate change, or toxic contamination of the environment, then he risks either being ignored or, worse, being subjected to ridicule. Such negative incentives reduce to a handful the number of scientists who are willing to speak out."[68] Thus, it is not reciprocity *for reciprocity's sake* that obliges our institutions to be more attentive to a wide range of expertise, nor is the latter simply that expertise which speeds up production of sought-after biomedical goods. Instead, an inclusive attention may produce better science in not weeding out individuals who may add valuable insights about the unintended risks or hidden possibilities of pursuing a particular course of action:

> To the extent that institutions become channels through which the talents and energies of the members of society can be expressed in

service to humanity, a sense of reciprocity will grow in which the individual supports and nurtures institutions and these, in turn, pay sincere attention to the voice of the people whose needs they serve. . . . Meeting this challenge implies a *fundamental change in the process of decision-making*, both individual and collective. Today, unbridled competition, obsession with power, and the abuse of authority vitiate the way decisions are made. The process suffers from extremes: apathy or over-enthusiasm, attachment to technique or haphazardness, devotion to minutia or the propensity to deal only with abstractions. What is vitally needed is a mode of operation into which systematic learning has been woven.[69]

By applying the principle of justice at every stage of decision-making, policymakers can "avoid the pitfalls of uniformity while still respecting the exigencies of equity."[70]

Recreating Decision-making Bodies as Learning Organizations. "What is at stake is the transformation of the present mode of governance, based on traditional concepts of power and authority, into one shaped by a genuine posture of learning . . . [to] foster in the inhabitants of each region the capacity to make increasingly more valid choices, both individually and collectively, regarding the development, transfer, and adoption of technology."[71] The social spaces necessary for this transformation must be widespread, embracing traditional learning centers such as universities, virtual realities such as the blogosphere, and public spaces such as town halls and community centers—but also settings we do not typically associate with collective deliberation over science and technology, such as health clinics, job training centers, places of worship, farms, and factories. In all these settings, a deeper awareness must be cultivated that science and technology are *not* neutral, "but at a more fundamental level, technology carries with it an ideology and pronounces on the way individual and social life should be organized. Technological choice bears on every other choice made about the quality and direction of life in a region. It is itself an expression of values."[72] The challenges to such bottom-up engagement cannot be overstated.

As some of the foremost analysts of participatory science have pointed out, "Science and scientists have not been used to the [social] context speaking back, so it is not surprising that they see contextualization as a challenge to their cognitive and social authority. . . . They fear that irrationality will break through the fragile crust of scientification. There is hard evidence that such fears are exaggerated."[73]

Forgetting to Remember

In *Autonomous Technology: Technics-out-of-Control as a Theme in Political Thought* (1978), Rensselaer Polytechnic Institute professor of political science Langdon Winner explores how many contemporary problems are due to "the widespread idea that technology is an autonomous force inducing change in society in ways beyond the control of human beings. Some celebrated the operation of this force, and others lamented it. But both groups were victims of the subtle paralysis of thought that such a belief produces in everyone who submits to it."[74] So the task of cultivating decision-making capacity and power-sharing structures cannot be directed simply at technophiles or at technophobes but at the seemingly passive middle as well. To that end, the passivity of Victor Frankenstein as depicted in *Autonomous Technology* provides a cautionary tale, for here is a person who "refuses to ponder the implications of his discovery":

> He is a man who creates something new in the world and then pours all of his energy into *an effort to forget*. His invention is incredibly powerful and represents a quantum jump in the performance capability of a certain kind of technology. Yet he sends it out into the world with no real concern for how best to include it in the human community. . . . He never moves beyond a dream of progress, the thirst for power, or the unquestioned belief that the products of science and technology are an unqualified blessing for humankind. Although he is aware of the fact that there is something extraordinary at large in the world, it takes a disaster to convince him that the responsibility is his. Unfortunately, by the time he overcomes his passivity, the consequences of his deeds have

become irreversible, and he finds himself totally helpless before an unchosen fate.[75]

From one perspective, the architects of Prop. 71 can be seen as breaking away from the passivity described above, instead seeking to direct the fate of science and technology toward the speedy production of novel biomedical goods. However, as the preceding chapters demonstrate, they did so within the narrow frame of consumptive biological citizenship and by implicitly excluding a number of vital concerns related to race, gender, class, and disability recognition and redistribution.[76]

In moving forward, it is vital that we construct participatory models for the development of science and technology by drawing upon the tools developed by practitioners experimenting with decision-making and power-sharing. Evolving over the past forty-five years, the Interactivity Foundation has been honing its approach to policy discussions on topics that run the gamut from crime and punishment to human impacts on climate change, the future of childhood and higher education, immigration, as well as those related to new biotechnologies. Likewise, Farzam Arbab's FUNDAEC approaches people as potential resources, not as problems to be managed nor as beneficiaries to be served. The Interactivity Foundation and FUNDAEC, among other organizations, avoid offering people "prepackaged" solutions on which they simply vote; rather, the process of *elaborating alternatives* is itself central to their method.

A number of analysts have drawn upon the classical idea of the "agora" as a civic platform or process to bring together different perspectives about science and technology. It is "the domain (in fact, many domains) in which contextualization occurs . . . neither state nor market, neither exclusively private nor exclusively public, the agora is the space in which societal and scientific problems are defined, and where what will be accepted as a 'solution' is being negotiated."[77] Some describe it as the emergence of a "new social contract" between science and society, wherein reliable knowledge must give way to socially robust knowledge, and scientific autonomy gives way to scientific accountability.[78] Others characterize a triple helix model of university, government, and industry relations, in which "the distinction between laissez faire and active-state intervention [is] obsolete," replaced by recombination of these three arenas, to which I would add a fourth, civic sphere that proves more challenging to integrate on an "equal level in the network."[79] In the actual implementation of this new social contract, public engagement has taken the form of onetime "consensus conferences," at which stakeholders gather to explore the various issues related to a particular development in science and technology. While this is certainly an improvement over the more exclusive and top-down forms of decision-making, a more radical shift is necessary still, in which engagement is more pervasive, ongoing, and informal, so that the capacity for thoughtful decision-making, and ultimately power-sharing, is not relegated to a participatory elite.

An example of cultivating decision-making power can be found at UCLA's Center for Society and Genetics' initiative at the King Drew High School in Watts, Los Angeles. As part of this program, students explored topics ranging from DNA forensics, pharmacogenomics, genetic discrimination, newborn screening, near-relative DNA forensic testing, gender testing, and designer babies, among other topics. Through debates, discussion questions, games, plays, videos, and writing exercises over the course of an academic year, students became familiar with the science behind these developments. But, and this is crucial, they also acquired skills to thoughtfully engage the ways in which science and technology acquire meaning for them personally, as well as for their families and communities.[80] Crucial to such initiatives is that participants consider a range of alternatives with respect to how various biotechnologies can be integrated in social life to bring about social equity.

While this and similar programs are currently configured as "science education outreach," in part to secure grants that are framed as such, in reality this form of capacity-building has as much to do with civic empowerment. On the surface, this type of program is not a radical departure from existing programs we typically find at elite prep schools at which students are provided with opportunities to build and experiment with science and technology in ways that prepare them to be drivers of innovation later in life. What distinguishes UCLA's program, in part, is that it engages young people from a predominantly African American and Latino working-class neighborhood, a demographic that comprises 99 percent of the student body.[81] Empowering young people early on, not only with the requisite scientific know-how but also with the skills to critically think about and discuss the social, economic, and political issues that are inherent in the innovation process, is perhaps one of the most important arenas for cultivating participatory science. It places the onus on educational institutions—and, and even more, on public and private funders—to recognize the importance of this broader skill set, which is less a matter of "community outreach" and "science education" than it is one of empowering young people, and the wider communities of which they are a part, to become producers of knowledge. This is not just scientific and technical knowledge but also ethical and sociopolitical knowledge that both complements and complicates conventional thinking about innovation as always progressive.

These efforts to cultivate participatory science are all the more pressing because of the way that science and technology are busily reconstituting peoples' biological as well as social lives. The bioconstitutional struggles we have discussed reveal the tension between the incredible value and commercialization of human tissue for research and the simultaneous dilemmas posed by human agency and interests. One study found that "the root of public ambivalence [about commercialization] seems to lie in (i) notions of justice and fairness about private profit being made through public exploitation, and (ii) a perceived lack of control in terms of governance."[82] For this reason, stem cell research and human genetics are spurring creative proposals for reconstituting the relationship between science and society so as to move us beyond the constitutional amendments and electoral politics that characterize Prop. 71 and a host of copycat legislation.

For example, David Winickoff proposes a legal architecture for

partnership governance [that] would empower participants to exert a share in distributive decision-making in return for contributing to the economic and social capital of the project. . . . Partnership governance seeks to go further than existing mechanisms of "community consultation," by implementing control rights at the level of the research participant collective. . . . [B]uilding an architecture *ex ante* for the legitimate representation of research participants, before challenges are encountered, is likely to be a good investment.[83]

Alluding to schemes like Prop. 71's royalty proposal, Winickoff points to the shortcomings of "benefit sharing [that] attempts to stitch a distributive norm at the seam of market and gift economies."[84]

By contrast, participatory science-making must repurpose existing initiatives: taking them apart, discarding features that stand in the way of civic inclusion, and redesigning them to be more in line with "partnership governance." At the same time, to transform the social fabric out of which future initiatives will be designed, we must continue to create more fundamental changes. Among these, empowering young people must be among our foremost priorities. As the most sought-after consumer market, they must likewise be at the forefront of casting off the expectation that they are simply consumers of scientific and technical goods produced by an innovative elite. Rather, they, as do we all, have both the right and responsibility to conceive new kinds of goods—material and social—that don't simply benefit us personally but contribute to our collective well-being as well.

ACKNOWLEDGMENTS

I have always found bylines a bit suspect, because of how they mystify who and what are actually involved in the completion of any project. What follows then is the equivalent of Dorothy pulling the curtain back after her long and storied journey through Oz—except that instead of finding a lone wizard pulling the strings (more mystification!), I am indebted to a host of relationships and webs of support in and outside of the academy for expanding my mind, invigorating my heart, and emboldening my nerve to undertake the work that led to this book.

Foremost among these are the many informants that tolerated my intrusions and engaged my queries over the course of this research. It is my hope that they find something useful in the preceding pages and that it touches upon the struggles and commitments they have graciously shared with me. I hope that at those points where I have not reached the same conclusions as they have, we can use that as a reason to open future engagement.

In the many iterations of this project, I have been inspired and challenged at every turn by the work and insights of Troy Duster, who opened my mind to the possibilities of investigating how authoritative knowledge grows out of particular historical and cultural conditions ("location, location, location," as he says). Amidst the frequent demonstrations in Sproul Plaza, Troy introduced me to a different mode of social change that challenges social structures at their core—where scientific reason appears autonomous and power seems inevitable. It was in working with him that I began putting together the analytical toolkit that I bring to bear here.

Were it not for Charis Thompson's suggestion to study what was then right in front of me, rather than running away to some distant and romantic field-work site, I am certain I would have missed the import of the public investment in stem cell research unfolding at the time. Thinking alongside Charis, whose razorlike incisiveness opens up the innards of any social phenomena, I eventually came to appreciate how my prior research on the cultural politics of child-birth shared many overlapping concerns with struggles over stem cell research. I am forever grateful to my guide from the "low-tech" world of midwifery to the "high-tech" world of biotechnology for introducing me, in a very understated way, to what ethnographers call "studying up"—examining the inner workings of powerful institutions and elite individuals as a useful route to questioning (and eventually undoing) the status quo.

Loïc Wacquant, more than anyone, kept me on my toes, holding up the punching pads to my thinking and writing, challenging me to become more precise, agile, and careful in my analysis of the social world. His insights were essential to helping me unwrap the shiny packaging of populism that sparked this inquiry and reexamine group spokesmanship with fresh eyes. Loïc, above all, kept his promise: "to push me beyond the limits that I set for myself," which is an approach to intellectual training that I continue to bring to my own teaching and advising.

I am grateful to many others at UC Berkeley: Michael Burawoy, Neil Fligstein, Cori Hayden, Denise Herd, Percy Hintzen, Elaine Kim, Samuel Lucas, Kristin Luker, Evelyn Nakano-Glenn, Trond Petersen, Raka Ray, Nancy Scheper-Hughes, Barrie Thorne, Ruth Wilson Gilmore, as well as the indefatigable Elsa Tranter and Gloria Chun. I also wish to extend my heartfelt appreciation to Spelman College faculty members—Barbara Carter, Harry Lefever, Mona Phillips, Cynthia Spence, Bruce Wade, and Darryl White—whose passion for social inquiry was so infectious that I could not resist. Likewise, Beverly Guy Sheftall and Bahati Kuumba were my first introduction to feminist theory; Opal Moore and Ray Grant kept my love of writing alive; and the late Roy Martinez's take on philosophy excited my ongoing commitment to dialogical thinking and teaching. Even earlier, at the United World College of Southern Africa (Waterford Kamhlaba), Patricia Earnshaw instilled a love and appreciation of history in me that I hope remains evident in my work.

At UCLA's Center for Society and Genetics, I benefited tremendously from a community of scholars who demonstrated that "interdisciplinarity" is much more than an academic buzzword. The center was a generative space to think *deeper* about my own disciplinary assumptions and *wider* across the academic boundaries that had become safe (and stifling) after a while. I thank directors Norton Wise and Edward McCabe for seeing promise in my work, and colleagues Soraya de Chadarevian, Debra Greenfield, Christopher Kelty, Hannah Landecker, Jessica Lynch Alfaro, John Novembre, Aaron Panofsky, Charles Taylor, as well as Laura Foster and Carrie Friese for being incredibly generous with their time and feedback. At UCLA, I was especially fortunate to work with Stefan Timmermans. In addition to the ease with which he would zoom out to provide framing guidance and then home in to discern the smaller inconsistencies that inevitably remained, I am still awed by how responsive he was to my many, often urgent, queries.

Colleagues-turned-friends who have, in multiple ways, managed to simultaneously encourage me to persevere in this scholarly marathon, provide me with much needed emotional water breaks, and remind me that *life is not a race* are many: Kemi Balogun, Laleh Behbehanian, Emily Brissette, Siri Colom, Dawn Dow, Shannon Gleeson, Jennifer Liu, Shabnam Khorala-Azad, Osagie Obasogie, Tianna Paschal, Tom Pessah, and all of my coworkers in the Sociology Diversity Working Group; my writing group, Catherine Bliss, Gwen D'Arcangelis, Brian Folk, Nat Turner, and Manuel Valle, for critical conversations and feedback and without whom I might still be holed up in a Berkeley café talking with an imaginary Bourdieu; during the final stages of writing, Aaron Panofsky, Sara Shostak, and especially Catherine Bliss pushed me, one "tracked change" at time, across the finish line.

Sheila Jasanoff at the Harvard Kennedy School of Government's Program on Science, Technology, and Society intervened at a critical juncture and provided valuable input that helped me pull together what was a more disjointed set of discussions. The Tarrytown Meetings, organized by Richard Hayes and Marcy Darnovsky with the Center for Genetics and Society, provided an incredibly supportive space to think about the implications of this research and make useful connections with those working on similar issues in and outside of the academy. The work of CGS's Jesse Reynolds and Emily Galpern and the California Stem Cell Report's David Jensen were also invaluable to my investigation.

I want to also recognize my colleagues at Boston University—an exceedingly supportive and brilliant mix of people who I am thrilled are willing to put up with me. Likewise, I have benefited from the critical feedback and encouragement of Rene Almeling, James Battle, Phillipe Copeland, Shawna Floyd, Duana Fullwiley, Matthew Hughey, Ben Hurlbut, Tabatha Jones Jolivet, L'Heureux Lewis-McCoy, Betram Lubin, Michael Montoya, Alondra Nelson, Judy Norsigian, Abena Osseo-Asare, Anne Pollock, Dorothy Roberts, Wendy Roth, Krishanu Saha, Patricia Torres, Patricia Wilson, and David Winickoff. I have also been extremely fortunate to be surrounded by a host of mentors without whose incredible example and guidance I would have surely lost sight of the forest for the trees: Beverlee and Richard Allen, Jamila Canady, Robin Chandler, Joy and Oscar DeGruy, Renee Dixon, Adalia Ellis, Amanda Enayati, Tod and Alison Ewing, Angela and Karim Ewing-Boyd, Nii-Odoi Glover, Ruth Forman and Jack Gordon, Vernada and Nasif Habeeb-ullah, Cynthia and Chuck Hall, Walter Heath, Stephanie and William Jeffries, Allison and Joseph Khoury, Ginny and Greg Kintz, Kimimila Locke, Jayne Mahboubi, Rose and Daniel McCoy, Solomon Moore, Asali Powers, Tadia Rice, Sara and Farhad Shadravan, William "Smitty" Smith, Munirih Taafaki, Leili Towfigh, Margene and Bill Willis, and many other Bahá'í coworkers who have encouraged me to pursue "those branches of knowledge which are of use" rather than those that "begin and end in words."[1]

Completion of this research would not have been possible without generous support from multiple sources: the National Science Foundation; Ford Foundation; California Institute for Regenerative Medicine; UC Berkeley Center for Race and Gender; UC Berkeley Center for Science, Technology, Medicine, and Society; UC Berkeley Townsend Center for the Humanities; UCLA Institute for Society and Genetics; Boston University Morris Fund; and most recently the American Council of Learned Societies, and Harvard University Kennedy School of Government's Program on Science, Technology, and Society.

My thanks go out to Kathy Jackson for her stellar transcription work, Shahrzad Abbassi-Rahbar for research assistance, Robert Cohen for extraordinary editorial assistance, and John Alan Birch, Kenneth Burney, Dustin Guinee, Joy Liu, and Bre Vader for their creative generosity. I also wish to extend my deepest appreciation to Kate Wahl and her team at Stanford University Press, whose combination of professionalism and humanity is matchless.

Last, and most of all, my *kin extraordinaire*: all the Mihrshahis and the Whites spread near and far; sisters and brothers "from other mothers," especially Nesanet Abegaze, Abena Agyeman-Fisher, Alisha Alexander, Allison Anastasio, L'Erin Asantewaa, Zhaleh Boyd, Pamela Douglas, Liz Dwyer, Iyabo Kwayana, Naimeh Lawrence, Laura Luna, Miatta Mansion, Sayeh Martin, Juliet Mathenge Jenkins, Akia Pacheco, Rubi Pacheco-Rivera, Claudia Pena, Ody Perez, Kamal Sinclair, Razi Khamsi Wilson, Tatiana Zamir, and all their significant others, whose love and encouragement know no bounds and who have brought endless joy, encouragement, and "laugh games" into my life.

To my beloved parents, Behin and Truitt White, brother Jamal, sons Malachi and Khalil, and *above all*, Shawn Benjamin ... there are really no words to express my gratitude for being buoyed up over so many years of research and writing. With support like this, I "never needed anything from the fake wizard, anyway."[2]

NOTES

Last Internet site access was Dec. 31, 2012, unless otherwise stated.

Preface

1. Bliss, Catherine. 2012. *Race Decoded: The Genomic Fight for Social Justice.* Stanford, CA: Stanford University Press.
2. I am indebted to Truitt White for sharing this set of insights with me.
3. Phelan, J. C., and Link, B. G. 2005. "Controlling Disease and Creating Disparities: A Fundamental Cause Perspective." *Journals of Gerontology 2*: 27–33.
4. Collins, P. H. 1986. "Learning from the Outsider Within: The Sociological Significance of Black Feminist Thought." *Social Problems 33* (6): S14–S32.
5. I acknowledge Sheila Jasanoff for introducing me to this turn of phrase.

Introduction

1. Personal interview, Aug. 21, 2007.
2. Sennett, R. 1980. *Authority.* New York: Norton, p. 196.
3. Gerardi Riordan, Donna. 2008. "Research Funding via Direct Democracy: Is It Good for Science?" *Issues in Science and Technology* (Summer), http://www.issues.org/24.4/p_riordan.html
4. Field note transcription. Feb. 16, 2007. San Francisco Sheraton, Burlingame, California.
5. Gibbons, M. 1999. "Science's New Social Contract with Society." *Nature 402*.
6. Reardon, J. 2004. *Race to the Finish: Identity and Governance in an Age of Genomics.* Princeton, NJ: Princeton University Press.
7. Washington, H. 2007. *Medical Apartheid: The Dark History of Medical Experimentation on Black Americans from Colonial Times to the Present.* New York: Doubleday; Reverby, Susan M. 2009. *Examining Tuskegee: The Infamous Syphilis Study and Its Legacy.* Chapel Hill: University of North Carolina Press.
8. Epstein, S. 1996. *Impure Science: AIDS, Activism, and the Politics of Knowledge.* Berkeley: University of California Press.

9. Shelton, Deborah L. 2010. "Who Owns Your Genes?" *Chicago Tribune*, Apr. 1.

10. Steenhuysen, Julie. 2011. "Court Upholds Patenting of Genes in Myriad Case," Reuters, July 29.

11. Shelton, "Who Owns Your Genes?"

12. Steenhuysen, "Court Upholds Patenting of Genes."

13. Bush, V. 1945. *Science The Endless Frontier.* Washington, DC: United States Government Printing Office. Available at: http://www.nsf.gov/od/lpa/nsf50/vbush1945.htm

14. Etzkowitz, H. 1998. "The Norms of Entrepreneurial Science: Cognitive Shifts in the New University-Industry Linkages." *Research Policy 27*: 823–33; Etzkowitz, H., and Leydesdorff, L. 1997. *Universities and the Global Knowledge Economy: A Triple Helix of University-Industry-Government Relations.* London: Cassell.

15. See Bliss, Catherine. 2012. *Race Decoded: The Genomic Fight for Social Justice.* Stanford, CA: Stanford University Press.

16. The twenty-nine-member governing board of the CIRM, or the Independent Citizens' Oversight Committee (ICOC), includes appointees by major political and academic officials such as the governor, the state controller, and the chancellor of the University of California; there are also ten preidentified disease representatives, as well as representatives of the biotech industry.

17. See Jasanoff, S., ed. 2011. *Reframing Rights: Bioconstitutionalism in the Genetic Age.* Cambridge, MA: MIT Press, for a full elaboration of the bioconstitutional framework and case studies.

18. Similarly, UC Berkeley anthropologist Cori Hayden explains that scientific knowledge "does not simply represent (in the sense of *depict*) 'nature,' but it also represents (in the *political sense*) the 'social interests' of the people and institutions that have become wrapped up in its production." See Hayden, C. 2003. *When Nature Goes Public: The Making and Unmaking of Bioprospecting in Mexico.* Princeton, NJ: Princeton University Press, p. 21.

19. Other states have begun to follow California's lead. In the 2006 midterm elections, stem cell research was expected to be an issue in at least thirty Senate, House, and gubernatorial campaigns, and was expected to play a significant role in at least eight of these thirty. As reported by the Center for Genetics and Society, stem cell research only met this expectation in four of the eight races. See Center for Genetics and Society. 2006. "Stem Cell Research in the Midterm Elections" (Nov. 8). Available at: http://www.geneticsandsociety.org/article.php?id=1965

20. Gottweis, H. 1998. *Governing Molecules: The Discursive Politics of Genetic Engineering in Europe and the United States.* Cambridge, MA: MIT Press; Wright, Susan. 1994. *Molecular Politics: Developing American and British Regulatory Policy for Genetic Engineering, 1972–1982.* Chicago: University of Chicago Press.

21. Personal interview with Joseph Tayag. Nov. 18, 2006. Berkeley, California.

22. Scholars refer to this interaction effect between science and society as "coproduction." For an elaboration, see Jasanoff, S., ed. 2006. *States of Knowledge: The Co-Production of Science and the Social Order*. New York: Routledge; Reardon, J. 2004. *Race to the Finish: Identity and Governance in an Age of Genomics*. Princeton, NJ: Princeton University Press.

23. Field note transcription. Oct. 26, 2006.

24. In fact, the suit *Mary Scott Doe et al. v. Robert Klein, et al.*, holds that the California stem cell amendment (Prop. 71) would "deprive Mary Scott Doe and other human embryos rights to due process and the equal protection of the laws in violation of Cal. Const. Art. I, §7(a), and will subject Mary Scott Doe and other human embryos to involuntary servitude and slavery in violation of Cal. Const. Art. I, §6."

25. Field note transcription. *Toward Fair Cures* conference, Oct. 14, 2006. Children's Hospital Oakland Research Institute, Oakland, California.

26. Epstein, S. 1996. *Impure Science: AIDS, Activism, and the Politics of Knowledge*. Berkeley: University of California Press, p. 350.

27. The significance of this discovery rests on the fact that human embryonic stem cells (ESCs) are "undifferentiated," meaning that they can potentially turn into any of the 220 types of cells and tissues in the human body. Through effective methods of culturing, these cells can be kept undifferentiated and subsequently used to study unknown aspects of biological development.

28. Scott, C. T. 2005. *Stem Cell Now: From the Experiment That Shook the World to the New Politics of Life*. New York: Pi Press, p. 9.

29. Jasanoff, S. 2005. *Designs on Nature: Science and Democracy in Europe and the United States*. Princeton, NJ: Princeton University Press, p. 6.

30. CBS News. 2010. "21st Century Snake Oil" (Apr. 16). Available at: http://www.cbsnews.com/2100-18560_162-6402854.html; Vogel, G. 2010. "Stopping Stem Cell Snake Oil" (June 22). Available at: http://news.sciencemag.org/scienceinsider/2010/06/stopping-stem-cell-snake-oil.html

31. Halper, E. 2008. "California's Budget Gap at $16 Billion" (Feb. 21). Available at: http://www.latimes.com/news/local/la-me-budget21feb21,0,6427050.story

32. "Snapshot: California's Uninsured 2007." Available at: http://www.policyarchive.org/handle/10207/5274

33. The top 5 percent of California earners average $207,363 annually, while the bottom 20 percent average $16,773. See Economic Policy Institute and Center on Budget and Policy Priorities. 2006. "Mind the Gap: Income Inequality, State by State." Available at: http://money.cnn.com/2006/01/25/news/economy/income_gap/

34. Available at: http://cpr.ca.gov/cpr_report/form_follows_function/chapter _1.html

35. Zhang, J., and Patel, N. 2005. *The Dynamics of California's Biotechnology Industry.* San Francisco: Public Policy Institute of California, p. 116.

36. Available at: http://www.cirm.ca.gov/press/pdf/2006/02-10-06.pdf

37. "Corporate income tax breaks would affect very few biotech firms because most of them have never made any profit. Thus, it is not an effective policy lever. In contrast, how to treat business losses has a great effect on the biotech industry. . . . Thus, some states allow biotech firms to sell their unused operating losses to other taxpayers" (Zhang and Patel 2005, p. 102).

38. Longaker, M. T., Baker, L. C., and Greely, H. T. 2007. "Proposition 71 and CIRM: Assessing the Return on Investment." *Nature Biotechnology 25* (5): 513–21.

39. 42 percent of voters make over $75,000 per year, while only 8 percent of voters make under $20,000.

40. Wacquant, L. 2004. "Pointers on Pierre Bourdieu and Democratic Politics." *Constellations 11* (1): 3–15.

41. Aug. 5, 2005. Regular ICOC Meeting. UC San Diego, La Jolla, California. Field note transcription at: http://www.cirm.ca.gov/files/transcripts/pdf/2005/08-05-05.pdf

42. Brown, M. B., and Guston, D. H. 2005. "Science, Democracy, and the Right to Research." *Science and Engineering Ethics 15* (3): 1471–1546.

43. By privileging the principle of autonomy, the stem cell initiative inherits a uniquely American bioethical framework in which "the socio-moral importance of persons, reciprocity, solidarity, and community are overshadowed by individualism" (Fox, R. 1990. "The Evolution of American Bioethics." In *Social Science Perspectives on Medical Ethics,* edited by G. Weisz. Boston: Kluwer Academic; cf. Myser, C. 2007. "White Normativity in United States Bioethics: A Call and Method for More Pluralist and Democratic Standards and Policies." In *The Ethics of Bioethics,* edited by L. Eckenwiler and F. Cohen. Baltimore: Johns Hopkins University Press, pp. 241–59, 244).

44. This dynamic is partially explored in a flourishing literature on biopolitics, which illustrates how modern governments exert and diffuse power precisely by controlling, interfering with, and propagating populations through the production of biological knowledge. See Foucault, M. 1978. *The History of Sexuality,* vol. 1. New York: Vintage; 1980. *Power/Knowledge: Selected Interviews and Other Writings.* New York: Pantheon.

45. Rose, N. 2007. *The Politics of Life Itself: Biomedicine, Power, and Subjectivity in the Twenty-First Century.* Princeton, NJ: Princeton University Press, p. 26. It is important to acknowledge that one of the first ground-breaking studies that theorized the concept focused squarely on how those affected by social, technological, and biological catastrophe petitioned the state for redress; see Petryna, A. 2002. *Life Exposed: Biological Citizens After Chernobyl.* Princeton, NJ: Princeton University Press.

46. Rose, N., and Miller, P. 1992. "Political Power Beyond the State: Problematics of Government." *British Journal of Sociology 43* (2): 173–205; Rabinow, P., and Rose, M. 2006. "Biopower Today." *BioSocieties 1*: 195–217.

47. "Biological citizens" are those who make demands upon the state for biomedical goods and articulate their agenda as a nonpolitical, consumer-based movement seeking scientific research that is unencumbered by social justice concerns that might slow or thwart their quest for cures. Reciprocally, these citizens are expected to take responsibility for their own health as a matter of civic duty, so that people and populations that are deemed "irresponsible" in this respect are what we might call "defectors," insofar as they do not perceive or engage their biological "defects" within the neoliberal frame of reference defined by the primacy of autonomy and self-enhancement, thereby also rejecting the terms of biological citizenship. See Rose, N., and Novas, C. 2005. "Biological Citizenship." In *Oikos and Anthropos: Blackwell Companion to Global Anthropology*, edited by A. Ong and S. Collier. Oxford, UK: Blackwell.

48. Good, M. D. 2001. "The Biotechnical Embrace." *Culture, Medicine and Psychiatry 25*: 395–410.

49. Insofar as "technologies of the self" presume an autonomous individual working on his or her body in a more or less private arena wherein the sovereign gaze of the state is displaced by myriad self-regulating activities, as in Foucauldian accounts, greater acknowledgment of the socioeconomic privilege associated with this consumptive and individualistic orientation is vital.

50. Hill Collins, P. 1990. *Black Feminist Thought: Knowledge, Consciousness, and the Politics of Empowerment.* New York: Routledge, p. 35.

51. For this reason, I resist the post-statist bent of much of the literature on biological citizenship and emphasize instead the coexistence of "old" and "new" assertions of power in my examination of contests in the stem cell arena. The centralized versus diffuse typology of power, as proffered in the Foucauldian framework, is not particularly useful in contexts where the role of the state is one of the main points of contention, and group interests are tied to whether power is exercised centrally via the authority of the state or diffusely via multiple corporate, organizational, and individual modalities.

52. Brown and Guston 2005.

53. Lee, O. 1998. "Culture and Democratic Theory: Toward a Theory of Symbolic Democracy." *Constellations 5* (4): 433–55, p. 445. For elaborations on how the effectiveness of particular collectives in the political sphere is directly linked to the categories "citizen" and "polity," and on the way in which these seemingly neutral categories are intersected by axes of social domination, see Fraser, N. 1992. "Rethinking the Public Sphere: A Contribution to the Critique of Actually Existing Democracy." In *Habermas and the Public Sphere*, edited by Craig Calhoun. Cambridge, MA: MIT Press; 1997. *Justice Interruptus: Critical Reflections on the "Postsocialist" Condition.* New York: Routledge; Calhoun, C.,

ed. 1992. *Habermas and the Public Sphere*. Cambridge, MA: MIT Press; Lee 1998; Young, Iris Marion. 1992. *Justice and the Politics of Difference*. Princeton, NJ: Princeton University Press.

54. Wacquant, L. 2006. *Punishing the Poor: The Neoliberal Government of Social Insecurity*. Durham, NC: Duke University Press.

55. Bourdieu's (1977) idea of political power incorporates the role of what he calls cultural "recognition" in constituting political power, and considers the ways in which political entrepreneurs make "representative claims" (Seward 2006) that also constitute the groups to which the claims are purportedly linked. He suggests that in the "discursive construction of the state," "[t]he same processes that enable one to construct the state also help one to imagine these other social groupings—citizens communities, social groups, coalitions, classes, interest groups, civil society, polity, ethnic groups, subnational groups, political parties, trade unions, and farmers organizations" (393). See Bourdieu, P. 1977. *Outline of a Theory of Practice*. Cambridge, UK: Cambridge University Press; Seward, M. 2006. "The Representative Claim." *Contemporary Political Theory* 5: 297–318.

56. The suits—one brought by a taxpayers group (the People's Advocate and the National Tax Limitation Foundation) and the other by a pro-life group (the California Family Bioethics Council, a project of the pro-life California Family Council)—charge that the institute violates the California state constitution because state oversight of its spending is inadequate.

57. Some private and state loans did eventually allow the institute to grant initial awards and fund the training program of which I was a part.

58. Institutional Review Board human subjects approval #2007-007.

59. Latour, Bruno. 1987. *Science in Action: How to Follow Scientists and Engineers Through Society.* Cambridge, MA: Harvard University Press, p. 99.

60. I acknowledge the assistance of Iyabo Kwayana and Dennis Hunter in helping with translation and clarifying the rich meaning enfolded within this phrase.

61. Meekosha, Helen, and Dowse, Leanne. 1997. "Enabling Citizenship: Gender, Disability and Citizenship in Australia." *Feminist Review* 57 (Autumn): 49–72, p. 50.

62. Meekosha and Dowse (1997) argue that theorization of private and public spheres has not encompassed the situation of people with disabilities, who "often inhabit a unique space that hovers stateless" (56).

63. For an elaboration of "technologies of humility," see Jasanoff, S. 2005. *Designs on Nature: Science and Democracy in Europe and the United States*. Princeton, NJ: Princeton University Press.

64. A recording of Watson making this statement while addressing a cohort of researchers can be viewed in the documentary film *Genomania*, by Paragon Media/Stephanie Welch.

Chapter One

1. Talk by Bernard Siegel, founder of the Genetics Policy Institute, speaking at the State of Stem Cell Advocacy 2008 conference, UCSF Mission Bay Campus, Apr. 12–13, 2008.

2. http://www.nationalreview.com/human-exceptionalism/326126/cirm -celebrates-funding-luxury-stem-cell-buildings-california-disintegra#

3. Wacquant, L. 2008. "Relocating Gentrification: The Working Class, Science and the State in Recent Urban Research." *International Journal of Urban and Regional Research 32* (1) (Mar.): 198–205.

4. Tayag, J. 2005. "The Color of Stem Cells: Why the Benefits of Stem Cell Research Might Not Be for People Like Me" (Sept. 9). Available at: http://www .geneticsandsociety.org/article.php?id=132

5. Moore, L. 2006. "Prop. 71, An Insider's Look from Illin n 'chillin.'" *Poor People's Magazine* (May 18). Available at: http://poormagazine.org/node/1630

6. Wacquant, L. 2004. "Pointers on Pierre Bourdieu and Democratic Politics." *Constellations 11* (1): 3–15. Available at: http://www.loicwacquant.net/assets /Papers/PB-POINTERSPBDEMOPOL.pdf

7. Whose Foods Coalition. 2011. "Open Statement: Whole Foods Must Sign a Community Benefits Agreement with Jamaica Plain" (Sept. 14). Available at: http://boston.indymedia.org/feature/display/213588/index.php

8. Gieryn, T. 1999. *Cultural Boundaries of Science.* Chicago: University of Chicago Press.

9. Brubaker, R. 2002. "Ethnicity Without Groups." *Archives of European Sociology 43* (2): 163–89, p. 167.

10. See Gottweis, H. 2008. "Participation and the New Governance of Life." *Biosocieties 3*: 265–86: "The 'new political spaces' opened up by Prop. 71 and other such initiatives are both inherently political and inherently social. They are *social* in that, as an ensemble of ideas and concepts, they are being 'produced, reproduced, and transformed in a particular set of practices' (Hajer 1995, 44). They are *political* in that such discourses do not merely disclose some underlying reality but 'actually constitute it' (Gottweis 2003b, 251). . . . The trend to experiment with new forms of participatory or 'interactive' policy making or 'interactive policy analysis' has been widely covered in the literature (e.g., Akkerman et al. 2004; Bellucci and Joss 2002; Chandler 2000). On a conceptual level, the literature has largely focused on developing taxonomies and systems of classification to assess and analyse the participatory practices that have been identified (e.g., Collins and Evans, 2002; Fiorino, 1990; Laird, 1993; see also Rowe and Frewer 2004). Central criteria in the assessment of these practices are those concerning *access* (who is allowed to participate; by whom is that determined) and *autonomy and influence* of the participants vis-à-vis formal political institutions. The exemplar case for the political sciences is the 'Ladder of Citizen Participation' by Arnstein (1969), which distinguishes

between eight levels of participation in political decision making, ranging from 'manipulation' to 'citizen control.' In the latter case, citizens fully control all stages of decision making, having been granted total sovereignty in reaching decisions."

11. Rose, N. 2007. *The Politics of Life Itself: Biomedicine, Power, and Subjectivity in the Twenty-First Century.* Princeton, NJ: Princeton University Press.

12. Ivri, I. 2005. "Cell Out: Management Issues Plague Distribution of $3 Billion in State Stem Cell Research Funds," Center for Genetics and Society (Mar. 24). Available at: http://www.geneticsandsociety.org/article.php?id=1614

13. May 6, 2005. Regular ICOC Meeting. Fresno Convention and Entertainment Center, Fresno, California. Field note transcription at: http://www.cirm.ca.gov/files/transcripts/pdf/2005/05-06-05.pdf

14. See Kerry A. Dolan. 2004. "The Top Ten U.S. Biotech Cluster." Available at: http://www.forbes.com/2004/06/07/cz_kd_0607biotechclusters.html (accessed July 7, 2011).

15. Bourdieu, P. 1985. "Social Space and the Genesis of Groups." *Theory and Society 14*: 723–44; 1989. "Social Space and Symbolic Power." *Sociological Theory 7* (1): 14–25. It may be more apt, however, to think of biopolitical struggles as a *battle*field rather than a *playing* field, since life chances are at stake.

16. Gottweis 2008.

17. May 6, 2005. Regular ICOC Meeting. Fresno Convention and Entertainment Center, Fresno, California. Field note transcription at: http://www.cirm.ca.gov/files/transcripts/pdf/2005/05-06-05.pdf

18. *Ibid.*

19. Little Hoover Commission. 2009. *Stem Cell Research: Strengthening Governance to Further the Voters' Mandate* (June), p. 2. Available at: http://www.lhc.ca.gov/studies/198/cirm/Report198.pdf

20. May 6, 2005. Regular ICOC Meeting. Fresno Convention and Entertainment Center, Fresno, CA. Field note transcription at: http://www.cirm.ca.gov/files/transcripts/pdf/2005/05-06-05.pdf

21. See "Q & A: Straight Talk from . . . Robert Klein." *Nature Medicine 13* (2007): 398.

22. As per sociologist Charis Thompson's work on "promissory capital," tissue economies grow based on "knowledge, technologies of life, and promise," and not simply on "production, productivity, and profit." Thompson, C. 2005. *Making Parents: The Ontological Choreography of Reproductive Technologies.* Cambridge, MA: MIT Press, p. 258.

23. Even so, in one of the agency's first decisions to disqualify a handful of applications that either missed the deadline or were incomplete, we observe the fragility of this legitimacy in the face of new stakeholders. That is, a number of criticisms were wielded at the agency in support of one or another site, most notably by board members and others who had obvious ties to the city in question.

24. From the text of Prop. 71, from the criteria for chairperson and vice chairperson of the ICOC (Independent Citizens' Oversight Committee), the designated governing body of the California Institute for Regenerative Medicine (CIRM): "Documented history in successful stem cell research advocacy. Experience with state and federal legislative processes that must include some experience with medical legislative approvals of standards and/or funding. . . . Legal experience with the legal review of proper governmental authority for the exercise of government agency or government institutional powers. Direct knowledge and experience in bond financing."

25. "Editorial: Stem Cell Czar?" *Sacramento Bee*, Dec. 16, 2004. Available at: http://www.geneticsandsociety.org/article.php?id=1553

26. Garvey, Megan. 2004. "Stem Cell Post Likely to Go to Klein." *Los Angeles Times*, Dec. 14. Available at: http://articles.latimes.com/2004/dec/14/local/me-stemcell14

27. Editorial. "Stem Cell Panel's Independence." *Lodi News-Sentinel*, Dec. 4, 2004.

28. Americans for Cures Foundation. n.d. "SB 1565 Action Alert for Appropriations Committee." Available at: http://www.americansforcures.org/article.php?uid=4939

29. John Simpson, stem cell policy director for Consumer Watchdog of Santa Monica, California, called for Klein's resignation, issuing this statement: "Don Gibbons, communications director for the California Institute for Regenerative Medicine, called this afternoon to tell me that Klein had stepped down from Americans for Cures. His phone call came after my posting the view today that holding both the state position and the advocacy position was untenable and the situation was a train wreck waiting to happen." July 14, 2008. See http://californiastemcellreport.blogspot.com/2008/07/klein-resigns-as-head-of-stem-cell.html

30. Americans for Cures Foundation. n.d. "Why We Oppose Senate Bill 1565." Available at: http://www.americansforcures.org/article.php?uid=4766

31. Field note transcription. Oct. 12–13, 2007. "The Stem Cell Meeting," hosted by Burrill and Company, UCSF Mission Bay Campus.

32. http://www.dailykos.com/story/2008/07/04/546462/—LAST-STAND-FOR-CALIFORNIA-STEM-CELL-PROGRAM (July 4, 2008).

33. Hall, C. T. 2005. "Stem Cells: The $3 Billion Bet: One Man's Scientific Mission; Housing Developer Leads California's Research Effort." Apr. 11. Available at: http://www.sfgate.com/cgi-bin/article.cgi?file=/c/a/2005/04/11/MNGKIC697P1.DTL

34. *Ibid.*

35. *Ibid.*

36. *Ibid.*

37. Lusvardi, W. 2011. "Cancel Prop. 71 Stem Cell Funding?" Feb. 23.

Available at: http://www.calwatchdog.com/2011/02/23/cancel-prop-71s-stem
-cell-research-funding/

38. Wacquant, L. 2008. "Relocating Gentrification: The Working Class, Science and the State in Recent Urban Research." *International Journal of Urban and Regional Research 32* (1): 198–205.

39. From the "Toward Fair Cures" conference publicity material. The conference had a mix of participants, some who were (and are) clearly stem cell "insiders," like the state's stem cell agency's then president, Zach Hall, and UC Berkeley's chancellor Robert J. Birgeneau, who is an ICOC board member. Other participants included well-known racial-ethnic minority advocates, such as the Greenlining Institute's then executive director (and cofounder), John Gamboa, and academics who deal critically with issues at the intersection of science, race, and class, such as medical historians Dr. Keith Wailoo and Dr. Evelynn Hammonds. The Greenlining Institute, which is located in Berkeley, California, and describes itself as "a multi-ethnic public policy research and advocacy institute . . . working for racial and economic justice," hosted the conference, which was cosponsored by UC Berkeley's Science, Technology and Society Center and the Children's Hospital Oakland Research Institute.

40. For a full account of the Black Panther Party's health activism, see Nelson, A. 2012. *Body and Soul: The Black Panther Party and the Fight Against Medical Discrimination.* Minneapolis, MN: University of Minnesota Press.

41. Field note transcription. Oct. 14, 2006. Talk at the Toward Fair Cures conference, Oakland, California.

42. *Ibid.*

43. *Ibid.*

44. Personal interview. Nov. 18, 2006. Berkeley, California.

45. Field note transcription. Oct. 14, 2006. Talk at the Toward Fair Cures conference, Oakland, California.

46. *Ibid.*

47. Field note transcription. Apr. 13, 2008. State of Stem Cell Advocacy conference, UCSF Mission Bay Campus.

48. *Ibid.*

49. Personal interview. Mar. 13, 2008. San Francisco. Sheehy also made the same point about HIV drug access in South Carolina at an IP Task Force Meeting, Nov. 22, 2005, at Stanford University.

50. "Yes on 71" and "No on 71" campaign contributions summary, see http://www.followthemoney.org/database/StateGlance/ballot.phtml?m=254.

51. Raw data used to calculate these figures can be found at http://cal-access.sos.ca.gov/Campaign/Committees/Detail.aspx?id=1260661&session=2003&view=expenditures.

52. Field note transcription. Apr. 13, 2008. State of Stem Cell Advocacy conference, UCSF Mission Bay Campus.

53. *Ibid.*

54. *Ibid.*

55. Personal interview. Feb. 2, 2007. San Francisco.

56. Personal interview. July 5, 2007. By telephone.

57. *Ibid.*

58. Rose, N., and Novas, C. 2005. "Biological Citizenship." In *Oikos and Anthropos: Blackwell Companion to Global Anthropology*, edited by A. Ong and S. Collier. Oxford, UK: Blackwell.

59. Roberts, D. 2012. *Fatal Invention: How Science, Politics, and Big Business Re-create Race in the Twenty-first Century.* New York: New Press, pp. 224–25.

60. Rose 2007.

61. Caplan, A. L. 2002. "Attack of the Anti-cloners." *The Nation*, June 17, pp. 5–6.

62. Fox, R. 1990. "The Evolution of American Bioethics." In *Social Science Perspectives on Medical Ethics*, edited by G. Weisz. Boston: Kluwer Academic, pp. 207–8. Cited from Myser, C. 2007. "White Normativity in United States Bioethics: A Call and Method for More Pluralist and Democratic Standards and Policies." In *The Ethics of Bioethics*, edited by L. Eckenwiler and F. Cohen. Baltimore: Johns Hopkins University Press, 241–59, p. 243.

63. Myser 2007, p. 248.

64. Fox 1990.

65. Wacquant, L. 2004. "Pointers on Pierre Bourdieu and Democratic Politics." *Constellations 11* (1): 11.

66. Little Hoover Commission 2009, pp. ii–iii, 2–3.

67. Reed, D. C. 2009. "The 'Do-Over' Commission: California Stem Cell Program to be Re-written?" (June 26). Available at: http://stemcellbattles.word press.com/2009/06/26/the-do-over-committee-california-stem-cell-program -to-be-re-written/

68. Jasanoff, S. 2006. "Transparency in Public Science: Purposes, Reasons, Limits." *Law and Contemporary Politics 29* (21): 21–39.

69. Don Reed's "Stem Cell Battles" blog. Available at: http://stemcell battles.wordpress.com/2009/06/26/the-do-over-committee-california-stem-cell -program-to-be-re-written/

70. This is what journalist David Jensen calls Reed and others who turn a blind eye to their own "faith-based" devotion to future cures: http://california stemcellreport.blogspot.com/2009/06/patient-advocate-slams-cirm-reforms-as .html

Chapter Two

1. Hagan, Pat. 2004. "Falling on Deaf Ears." *New Scientist 183* (2462): 36. Available at: http://business.highbeam.com/137753/article-1G1-121807315/falling -deaf-ears-far-being-grateful-some-deaf-people

2. Wolbring, G. 2003. "Confined to Your Legs." In *?Living with the Genie: Essays on Technology and the Quest for Human Mastery?*, edited by Alan Lightman, Daniel Sarewitz, and Christina Desser. Washington, DC: Island Press.

3. Field note transcription. Mar. 12, 2008. Crest Theater, Sacramento, California.

4. *Ibid.*

5. "Duchesneau gave birth to a boy, Gauvin, who was profoundly deaf in his left ear and had only limited hearing in his right. The couple rejected the offer of a hearing aid. It was a bizarre twist to the concept of designer babies. Usually, the selection process is about screening *out* disabilities—not creating them. Yet is the outcome so different to what goes on naturally when disabled people marry someone who has the same condition as them?" (Hagan 2004, 36)

6. *Ibid.*

7. Wolbring, G. 2008. "The Politics of Ableism." *Dialogue 51*: 252–58.

8. Kokonut Pundit. 2010. "Deaf Group Attacks Stem Cell Researcher for His Work on the Cure for Deafness" (Feb. 19). Available at: http://kokonutpundits.blogspot.com/2010/02/deaf-group-attacks-stem-cell-researcher.html

9. Despite the heterogeneity in types of ailments, organizational practices, and specific aims across patient advocacy groups, scholars provide evidence of shared values and discourse across organizations as indicative of a stem cell health movement. See Allsop, J., Jones, K., and Baggott, R. 2004. "Health Consumer Groups in the UK: A New Social Movement?" *Sociology of Health and Illness 26* (6): 737–56. This movement seeks to promote the interests of patients and their families and is increasingly characterized by heightened collaboration across organizations and even across disease types. Certainly, on one level, this movement can be understood as indicative of a shift in the direction of greater lay involvement in science and medicine. See Bastian, H. 1998. "Speaking Up for Ourselves." *International Journal of Technological Assessment in Health Care 14* (1): 3–23; Epstein, S. 1996. *Impure Science: AIDS, Activism, and the Politics of Knowledge*. Berkeley: University of California Press. Whether working on distinct organizational agendas or collaborating on a broader political agenda (e.g., stem cell research advocacy), disease advocates are part of a larger trend reflecting the predominance of interest group politics as a vehicle or proxy for "democratic inclusion." Given the nascent science involved in this case, however, the work of advocates in this arena is not characterized by a focus on any one illness or treatment but by a discourse highlighting the urgency to produce cures *in general*. This shift from seeking medical advances mainly through affiliation with a specific organization and on behalf of its members to making broad claims on behalf of ill and impaired people in general is what leads its participants to call this a "pro-cures movement."

10. Patient advocates are individuals and organizations working to address the needs and advance the interests of people affected by particular ailments.

They have become a prominent feature in the development of scientific and biomedical research agendas in recent years, spawning thousands of disease-specific organizations, lobbying governments (Heath, Rapp, Taussig 2004; Rose and Novas 2005), partnering with private companies (Ganchoff 2004), raising funds for research (Taussig 2005), challenging scientists' research priorities (Epstein 2004), and forming support and self-help groups (Novas 2006; Rabinow 1999). Some disease advocates have even gone so far as to create tissue registries for scientists to use in research (Heath, Rapp, and Taussig 2004; Sundar Rajan 2006), while others have engaged in widespread community screening initiatives that have proven to substantially reduce the incidence of genetic disease among particular ethnic groups (Wailoo and Pemberton 2006). Heath, D., Rapp, R., and Taussig, K. 2004. "Genetic Citizenship." In *A Companion to the Anthropology of Politics*, edited by D. Nugent and J. Vincent. Oxford, UK: Blackwell; Rose, N., and Novas, C. 2005. "Biological Citizenship." In *Oikos and Anthropos: Blackwell Companion to Global Anthropology*, edited by A. Ong and S. Collier. Oxford, UK: Blackwell; Ganchoff, C. 2004. "Regenerating Movements: Embryonic Stem Cells and the Politics of Potentiality." *Sociology of Health and Illness 26* (6): 757–74; Taussig, K. S. 2005. "The Molecular Revolution in Medicine: Promise, Reality, and Social Organization." In *Complexities: Beyond Nature and Nurture*, edited by S. McKinnon and S. Silverman. Chicago: University of Chicago Press, pp. 223–47; Epstein, S. 2004. "Bodily Differences and Collective Identities: The Politics of Gender and Race in Biomedical Research in the United States." *Body and Society 10* (2–3): 183–203; Novas, C. 2006. "The Political Economy of Hope: Patient Organizations, Science and Biovalue." *Biosocieties 1*: 289–306; Rabinow, P. 1999. *Essays on the Anthropology of Reason.* Princeton, NJ: Princeton University Press; Sundar Rajan, K. 2006. *Biocapital: The Constitution of Postgenomic Life.* Durham, NC: Duke University Press; Wailoo, K., and Pemberton, S. 2006. *The Troubled Dream of Genetic Medicine.* Baltimore: Johns Hopkins University Press.

11. Asch, A. 2001. "Critical Race Theory, Feminism, and Disability: Reflections on Social Justice and Personal Identity." *Ohio State Law Journal 62*: 7; Crow, L. 1996. "Including All Our Lives: Renewing the Social Model of Disability." In *Exploring the Divide*, edited by C. Barnes and G. Mercer. Leeds, UK: Disability Press, pp. 55–72; Shakespeare, T. 1996. "Disability, Identity, and Difference." In Barnes and Mercer (1996, 94–113); Wendell, S. 2001. "Unhealthy Disabled: Treating Chronic Illness as Disability." *Hypatia 16* (4).

12. I later learned that from 1912 to 1916 Fremont was the earliest home of California's motion picture industry.

13. "The Roman Reed Bill (AB750) was signed in September 2000 and initially dedicated $1 million per year over 5 years for spinal cord injury research in the State of California. The amount of this fund was increased in 2001–2 to $2 million per year. The Roman Reed program was due to sunset in 2005,

but on September 9, 2004, Governor Schwarzenegger signed the renewal of the Roman Reed Spinal Cord Injury Research Act, AB1794, extending funding through January 1, 2011. Since that time, approximately $1.5 million per year has supported the most promising scientific research in neural regeneration. The Roman Reed Research funds are allocated to the University of California (UC), with the UC Office of the President appointing the management of the grants to the Reeve-Irvine Research Center." Available at: http://www.reeve.uci .edu/roman-reed-program.html

14. Personal interview. May 9, 2007. Fremont, California.

15. Don C. Reed. Oct. 2005. Available at: http://www.stemcellbattles.com

16. Wolbring, G. 2003. "Confined to Your Legs." In *?Living with the Genie: Essays on Technology and the Quest for Human Mastery?*, edited by Alan Lightman, Daniel Sarewitz, and Christina Desser. Washington, DC: Island Press, pp. 139–56.

17. Personal interview. May 9, 2007. Fremont, California.

18. Crow 1996.

19. Personal interview. May 9, 2007. Fremont, California.

20. As Shakespeare (1996) explains, this assumes an asymmetrical logic: "A situation where disabled people are defined by their physicality can only be sustained in a situation where non-disabled people have denied their own physicality . . . rather than interrogating the other, rather let us deconstruct the normality-which-is-to-be-assumed" (3).

21. Personal interview. May 9, 2007.

22. Crow 1996.

23. Wolbring 2003, 150.

24. Mar. 1, 2005. Regular ICOC Meeting. Stanford University. Field note transcription at: http://www.cirm.ca.gov/files/transcripts/pdf/2005/03-01-05.pdf

25. In one synthesizing article, Ferguson and colleagues explain that "[p]erhaps the most universally agreed upon point is that family responses to disability are immensely variable. . . . There is an increasingly dominant body of research that finds aggregate patterns of overall adjustment and wellbeing across groups of families with and without children with disabilities" (86). Ferguson, P. M., et al. 2000. "The Experience of Disability in Families: A Synthesis of Research and Parent Narratives." In *Prenatal Testing and Disability Rights*, edited by Erik Parens and Adrienne Asch. Washington, DC: Georgetown University Press.

26. Gordon, P. H., Feldman, D., and Crose, R. 1998. "The Meaning of Disability: How Women with Chronic Illness View Their Experiences." *Journal of Rehabilitation 64* (3): 5–11; Rubin, R. M., and White-Means, S. I. 2001. "Race, Disability and Assistive Devices: Socio-demographics or Discrimination." *International Journal of Social Economics 28* (10–12): 927–41. For example, African Americans are found to have a lower rate of employment following spinal cord injury (SCI), even when controlling for education. Whites with

SCI are 2.8 times more likely to be employed than Blacks. See Hess, D. W., Ripley, D. L., McKinley, W. O., and Tewsbury, M. 2000. "Predictors for Return to Work After Spinal Cord Injury: A 3-year Multicenter Analysis." *Archives of Physical Medicine and Rehabilitation 81* (3): 359–63; Krause, J. S., and Anson, C. A. 1997. "Adjustment After Spinal Cord Injury: Relationship to Gender and Race." *Rehabilitation Psychology 42* (1): 31–46.

27. Morgan, D.H.J. 1992. *Discovering Men.* London: Routledge.

28. The juxtaposition between a prior life of hyperphysicality and a present condition of embodied captivity is nowhere more pronounced than in the life of the late stem cell spokesman Christopher Reeve. Possibly even more consequential for the stem cell movement than Reeve's own rhetoric and effective use of the media to disseminate his pro-cures message will prove to be the way that Reeve, as a symbol of single-minded persistence, energized other advocates, even after his passing. Gray explains that "using his special status as the crippled Superman and his considerable charisma, Reeve has catalyzed the unification of most of the patient groups in the United States who focus on paraplegia and other spinal cord injuries" (1). Gray, G. H. 2001. *Cyborg Citizen: Politics in the Posthuman Age.* London: Routledge.

29. Lane, H., Pillad, R. C., and Hedberg, U. 2011. *People of the Eye: Deaf Ethnicity and Ancestry.* Oxford, UK: Oxford University Press.

30. Personal interview. Aug. 21, 2007.

31. *Ibid.*

32. In our conversation, Patty took up the broader politics of naming and applied it to gender categories: "I don't know if you've done a lot of . . . gender analysis, but, you know, a lot of people would say for trans[sexual] people, the problem is you don't know what gender you are, and you need to pick, and the problem is . . . you know, you need to fix your genitals, right? Whereas the pushback and, I would say, the more accurate framing really is that no, the problem is that there's a gender binary and that you only understand that there's a boy or a girl when in fact they're a full spectrum. And the social model of disability argues that, in fact, . . . people have always been disabled, that they are always going to be disabled, that there's a range of abilities and that it's fine."

33. Hacking, I. 1986. "Making Up People." In *Reconstructing Individualism: Autonomy, Individuality, and the Self in Western Thought,* edited by T. C. Heller, M. N. Sosna, and D. E. Wellbery. Stanford, CA: Stanford University Press, p. 259.

34. Personal interview. Aug. 21, 2007.

35. *Ibid.*

36. This corresponds to a statement by the British Council on Disabled People, in one of its training manuals related to stem cell research, that "the 'suffering' of disabled people has been used, much as it has been used by charities, to take forward a commercial and scientific agenda. To raise funds for the research and get the necessary regulation to carry it out, disabled people have

been presented as the objects of pity in desperate need of being cured." British Council on Disabled People. 2004. *Disability and Bioethics Resource Pack*, vol. 1. (Available on request.) "The very idea of 'curing' disability is the core element in the discrimination of disabled people because the 'curing ideal' resides in conformity and normalcy." Reindal, S.M. 2000. "Disability, gene therapy and eugenics—challenge to John Harris." *Journal of Medical Ethics* 26(2):89–94.

37. Heath, D., Rapp, R., and Taussig, D-S. 2004. "Genetic Citizenship." In *A Companion to the Anthropology of Politics*, edited by D. Nugent and J. Vincent. Oxford, UK: Blackwell.

38. May 23, 2005. Regular ICOC Meeting. The Tech Museum of Innovation, San Jose, California. Field note transcription at: http://www.cirm.ca.gov/files/transcripts/pdf/2005/05-23-05.pdf

39. Rose, N. 2007. *The Politics of Life Itself: Biomedicine, Power, and Subjectivity in the Twenty-First Century*. Princeton (NJ) and Oxford (UK): Princeton University Press; Sundar Rajan, K. 2006. *Biocapital: The Constitution of Postgenomic Life*. Durham, NC: Duke University Press.

40. Following Rabinow and Bennett (2007), "The question of what constitutes a good life today, and the contribution of the bio-sciences to that form of life must be vigilantly posed and re-posed" (5). In "Bioethics to Human Practice," working paper available at: http://anthropos-lab.net/wp/publications/2007/08/workingpaper011.pdf

41. Foucault, M. 1994. The Order of Things: An Archaeology of the Human Sciences. New York: Vintage.

42. I leave for another discussion the growing human enhancement movement, which seeks to create "super-normals" with enhanced abilities (hearing, sight, memory, etc.) that will make the normal range of ability we observe today appear inadequate.

43. Available at: http://whoeverfightsmonsters-nhuthnance.blogspot.com/2008/08/gregor-wolbring.html. For elaboration, see Wolbring, G. 2007. "Nano-Engagement: Some Critical Issues." *Journal of Health and Development (India) 3* (1–2): 9–29.

44. Heath, Rapp, and Taussig 2004.

45. 125290.20 ICOC Membership; Appointments; Terms of Office (a)(3), p. 2. While the designation "advocate" is commonly thought to refer to someone who represents the interests of a political constituency or social group, ICOC disease advocates are also almost all formally affiliated with research institutions that will ultimately be recipients of California stem cell research grants.

46. Talk by stem cell advocate Bernard Siegel, founder of the Genetics Policy Institute, speaking at the "State of Stem Cell Advocacy" conference, UCSF Mission Bay, Apr. 12–13, 2008. He went on to advise on ways for those in the movement in other states to create a positive legal climate for the advancement

of stem cell research similar to that which enabled Prop. 71, noting that his organization was poised to provide policy experts with talking points "to educate the public and media on stem cell issues."

47. Skocpol, T. 2004. "Voice and Inequality: The Transformation of American Civic Democracy." *Perspectives on Politics 2* (1): 3–20. In Bourdieuian terms, the disability perspective can be seen as disavowing the "doxa" of biotechnology and the pro-cures movement, and as a result is excluded from this field of struggle. By contrast, even when patient advocates appear critical of the pace or quality of biomedical advances, they still share the "common sense" of this biomedical-dominated, market-driven arena. See Bourdieu, P. 1985. "Social Space and the Genesis of Groups." *Theory and Society 14*: 723–44; 1989. "Social Space and Symbolic Power." *Sociological Theory 7* (1): 14–25.

48. Skocpol, T. 2004. "Voice and Inequality: The Transformation of American Civic Democracy." *Perspectives on Politics 2* (1): 3–20.

49. For example, one bioethicist who served on the Human Genome Project's ethical advisory committee before being vetted by the California agency for a position on one of its working groups said she felt like she had to prove her stem cell loyalty prior to being appointed: "For the first time having long been an advocate of doing stem cell research, to be asked to pass a litmus test—I knew I was in politics then—to pass a litmus test about what my views are before I can engage in a project on stem cell research . . . whether we like it or not we are totally enmeshed in politics." But, we might add, this is a kind of politics in which seemingly apolitical patient advocates are privileged among all other constituencies.

50. Hagan 2004, 36–39.

51. Available at: http://deafcapital.blog.com/2011/07/07/deaf-people-and-the-cyborg/

52. As Heath, Rapp, and Taussig (2004) explain, "Health advocates with chronic, life-threatening illnesses have had to face the challenge of crafting complex political-economic relations with the state and market in the quest for medical treatment, social services, and appropriate biomedical research. In this process, they have articulated demands for insider status in scientific controversies, and claimed credit for contributing to scientific advances. In their coalitional work, a generative mix of public and private resources has been assembled in the service of new citizenship claims." (160)

53. May 23, 2005. Regular ICOC Meeting. The Tech Museum of Innovation, San Jose, California. Field note transcription at: http://www.cirm.ca.gov/files/transcripts/pdf/2005/05-23-05.pdf

54. Consider, for example, this compelling appeal by the "grassroots umbrella" organization called Stem Cell Action Network (SCAN): "The Pro-Cures movement is comprised of hundreds of organizations representing millions of patients and their families. That is both its strength and its weakness. . . . Let's also be clear as to what SCAN is not. It is not the advocacy representative of any

singular disease group. It is not about representing the interests of Parkinson's patients over diabetics. Nor is it meant to supplant any other organization or usurp their activities. To the contrary, SCAN is the hub that unites these groups, that unifies their collective voices and serves to address the stem cell side of their respective communities." Eisen, Jeff. 2006. "Organize, Organize, Organize." Dec. 21. Available at: www.stemcellpage.com/index_files/Editorial122106.htm

55. Personal interview. Aug. 21, 2007.

56. Quigley, R. B. 2006. "Advocacy and Community: Conflicts of Interest in Public Health Research." *Public Health Policy and Ethics 19*: 223–35.

57. Landers, S. H., and Sehgal, A. R. 2004. "Health Care Lobbying in the United States." *American Journal of Medicine 116* (7): 474–77.

58. See Landers and Sehgal 2004: "Health care lobbying expenditures totaled $237 million in 2000. These expenditures accounted for 15% of all federal lobbying and were larger than the lobbying expenditures of every other sector, including agriculture, communications, and defense. A total of 1192 organizations were involved in health care lobbying. Pharmaceutical and health product companies spent the most ($96 million), followed by physicians and other health professionals ($46 million). Disease advocacy and public health organizations spent $12 million. From 1997 to 2000, lobbying expenditures by physicians and other health professionals grew more slowly than lobbying by other organizations (10% vs. 26%)." (474)

59. Cited in Lehrman, S. 2005. "A Proposition for Stem Cells" (Aug. 22). Available at: http://www.sciam.com/article.cfm?id=a-proposition-for-stem-ce& page=2 (accessed Dec. 28, 2012).

60. The social justice beliefs that inform the disability rights movement necessarily undermine its ability to become stem cell "insiders." Arenas such as those generated by the California stem cell initiative are implicitly antagonistic to political advocacy based upon forms of "modern" identity groupings such as race and gender, which disability advocates tend to emulate. Rather, the mode of "biosocial" group-making in which people organize and affiliate around a particular illness experience tends to ensure that patient advocates gain insider status, partly because they add moral valence to the interests of scientists and venture capitalists. Most importantly, biosocial group-making and related advocacy focused on personal experiences and free market choices are amenable to an apolitical framing that, paradoxically, grants them greater political muscle in controversial scientific arenas.

61. Goggin, G., and Newell, C. 2005. "Media Bios: Or *Harvie Krumpet* and the Ethics of Disability and Death." Paper presented at the Annual Meeting of the Australian and New Zealand Communication Association. Christchurch, New Zealand. July 4–7. Available on request.

62. Emphasis added. National Council on Disability (NCD), cited in Wolbring, Gregor. 2009. "Culture of Neglect: Climate Discourse and Disabled

People." *Media Culture Journal 12* (4). The NCD estimated that a dispropor-tionate number of Hurricane Katrina fatalities were people with disabilities: "most of those individuals had medical conditions and functional or sensory disabilities that made them more vulnerable."

Chapter Three

1. Field note transcription. Sept. 28, 2006. Ethical Worlds of Stem Cell Medicine conference. UC Berkeley, Townsend Center. Talk by Radhika Rao entitled, "Coercion, Commercialization, and Commodification: The Ethics of Compensation for Egg Donors in Stem Cell Research."

2. See www.ifrr-registry.org for more information. I wish to acknowledge Judy Norsigian for drawing my attention to this important new resource.

3. Center for Bioethics and Culture Network. 2011. "Market Competition Collision: Eggs Needed for Research" (interview, May 4). Available at: http://www.cbc-network.org/2011/05/market-competition-collision-eggs-needed-for-research/

4. *Ibid.*

5. Almeling, R. 2007. "Selling Genes, Selling Gender: Comparing Egg and Sperm Donors." *American Sociological Review 72*: 319–40; Beeson, D., and Lippman, A. 2006. "Egg Harvesting for Stem Cell Research: Medical Risks and Ethical Problems." *Reproductive BioMedicine Online 13* (4): 573–79; Foohey, P. 2010. "Paying Women for Their Eggs for Use in Stem Cell Research." *Pace Law Review 30*: 900–926.

6. See Ikemoto, Lisa. 2009. "Eggs as Capital: Human Egg Procurement in the Fertility Industry and the Stem Cell Research Enterprise." *Signs: Journal of Women in Culture and Society 34*, 763; Reynolds, Jesse. 2008. "The New Push for Eggs for Stem Cell Research in California." *Biopolitical Times* (posted June 27). Available at: http://www.biopoliticaltimes.org/article.php?id=4149. Franklin, S. 2006. "Embryonic Economies: The Double Reproductive Value of Cells." *Biosocieties 1*: 71–90.

7. The technique gained notoriety when researchers in the United King-dom used SCNT to clone Dolly the sheep, arousing worries that rogue re-searchers would use it to clone humans.

8. Available at: http://www.washingtonpost.com/wp-dyn/content/article/2009/06/25/AR2009062501931_2.html

9. See Robeznieks, A. 2003. "Researchers Ponder Best Use of 400,000 Stored Embryos." *AMNews*, June 16; Wade, N. 2003. "Clinics Hold More Em-bryos Than Had Been Thought." *New York Times*, May 9, p. A-24; Hoffman, D. I., Zellman, G. L., Fair, C. C., Mayer, J. F., Zeitz, J. G., Gibbons, W. E., and Turner, T. G., Jr., Society for Assisted Reproduction Technology (SART) and RAND. 2003. "Cryopreserved Embryos in the United States and Their Avail-ability for Research." *Fertility and Sterility 79* (5): 1063–69.

10. Gupta, J. A. 2011. "Exploring Appropriation of 'Surplus' Ova and Embryos in Indian IVF Clinics." *New Genetics and Society 30* (2): 167–80.

11. Maher, B. 2008. "Egg Shortage Hits Race to Clone Human Stem Cells." *Nature 453*: 828–29. Available at: http://www.nature.com/news/2008/080611/full/453828a.html

12. Implicit in Eggan's plea is a growing anxiety around what we may call the "bioscarcity" of eggs within the tissue economy. The term, referring to a variation of conventional economic scarcity, draws attention to the way that tissue suppliers can generate artificial shortages by refusing to participate in research initiatives.

13. While science and technology studies have trained us to consider the nonhuman inputs and agencies associated with the technological shift in biomedicine over the last thirty years, in Eggan's frustration we observe that techniques of recruiting stratified human populations into the research pipeline can make or break an initiative.

14. Romney, Lee. Sept. 17, 2006. Available at: Available at: http://articles.latimes.com/2006/sep/13/science/sci-eggs13.

15. Judy Norsigian of the organization Our Bodies, Ourselves offers a full list of problems: rash, vasodilation (dilation of blood vessels causing a "hot flash"), paresthesia (sensation of burning), tingling, pruritis (itching), headache and migraine, dizziness, urticaria (hives), chest pain, nausea, depression, emotional instability, loss of libido (sex drive), amblyopia (dimness of vision), syncope (fainting), asthenia (weakness), asthenia gravis hypophyseogenea (severe weakness due to loss of pituitary function), amnesia (disturbance in memory), hypertension (high arterial blood pressure), tachycardia (rapid beating of the heart), muscular pain, bone pain, nausea/vomiting, asthma, abdominal pain, insomnia, swelling of hands, general edema, chronic enlargement of the thyroid, liver function abnormality, vision abnormality, anxiety, myasthenia (muscle weakness), and vertigo.

16. Norsigian, J. 2005. "Egg Donation for IVF and Stem Cell Research: Time to Weigh the Risks to Women's Health." *Different Takes* (Spring).

17. *Ibid.*

18. "Although Lupron's 'off-label' use is allowed, and despite the fact that since 2006 the FDA has received more than 6,000 complaints regarding the drug, including twenty-five deaths related to 'off-label' use, little research exists about the exact short- and long-term effects of its use in connection with egg extraction" (Foohey 2010, 907).

19. While efforts to establish a moratorium on research have not been successful, it is partly in response to such efforts that the California stem cell agency began to actively determine the effects of egg procurement on women's health. For example, it joined with the Institutes of Medicine to cosponsor a "Public Workshop on Assessing the Medical Risks of Donation" in September 2006.

20. In stating this, Prop. 71 engineers chose to adopt the National Academy of Sciences' guidelines as an interim standard until it was able to institute its own detailed protocols, which turned out to be stricter than the National Academy of Sciences' standards.

21. Vesely, R. 2004. "California's Prop 71 Divides Debate on Stem Cells." *WeNews*, Oct. 26. Available at: http://womensenews.org/story/campaign -trail/041026/californias-prop-71-divides-debate-stem-cells

22. National Academy of Sciences' Committee on Guidelines for Human Embryonic Stem Cell Research, National Research Council. 2005. *Guidelines for Human Embryonic Stem Cell Research.* Washington, DC: National Academies Press, p. 84.

23. This is Senate Bill 1260 submitted by Senators Ortiz and Runner in 2006.

24. For the full text of the bill, see: http://www.cdph.ca.gov/services/ boards/HSCR/Documents/MO-SB1260-08-2007.pdf

25. Darnovsky, M. 2006. "A Pro-Woman Stem-Cell Policy." Nov. 16. Available at: http://www.tompaine.com/articles/2006/10/26/a_prowoman_stem cell_policy.php

26. Waldby and Mitchell turn our attention to the legal backdrop of the gift/commodity distinction, explaining that "[i]t reflects the English common-law principle that persons do not have a property right in their bodies, and hence cannot sell themselves or purchase another. . . . While donors are largely excluded in U.S. and British law from selling their tissues (with the exception of reproductive tissues and plasma in the United States), their donated tissue may be either sold by the receiving party (hospitals routinely sell tissues to pharmaceutical or cosmetics companies, for example) or transformed into cell lines or gene sequences and patented." Waldby, C., and Mitchell, R. 2006. *Tissue Economies: Blood, Organs, and Cell Lines in Late Capitalism.* Durham, NC: Duke University Press, p. 23.

27. Anderson, B. 1991 (new edition 2006). *Imagined Community: Reflections on the Origin and Spread of Nationalism.* London and New York: Verso.

28. Frow aptly explains that "there is no single form of 'the gift,' and no pure type of either the gift economy or the commodity economy. . . . the gift therefore cannot and should not be conceived as an ethical category: it embodies no general principle of creativity, of generosity, of gratuitous reciprocality, or of sacrifice or loss." See Frow, J. 1997. *Time and Commodity Culture: Essays in Cultural Theory and Postmodernity.* Oxford, UK: Oxford, p. 124.

29. Field note transcription. Sept. 28, 2006. Ethical Worlds of Stem Cell Medicine conference, UC Berkeley, Townsend Center. Talk by Radhika Rao entitled, "Coercion, Commercialization, and Commodification: The Ethics of Compensation for Egg Donors in Stem Cell Research."

30. Thompson, C. 2007. "Why We Should, in Fact, Pay for Egg Donation." *Regenerative Medicine 2* (2): 203–9.

31. Goodwin, M. 2007. "Private Ordering and Intimate Spaces: Why the Ability to Negotiate Is Non-negotiable." *Michigan Law Review 105*: 1367–86, 1369.

32. Dickenson, D. 2006. "The Lady Vanishes: What's Missing from the Stem Cell Debate." *Bioethical Inquiry 3*: 43–54, p. 53.

33. A related point is that the two feminist camps map onto a larger political discourse in which both sides espouse "women's interests" but come to their positions via different political and ethical diagnoses. The distinction between those seeking a moratorium and those seeking reciprocal exchange corresponds to different feminist intellectual traditions that have tended to highlight either oppressive social structures (i.e., structuralists) or the multiple ways actors engage with their social worlds (i.e., poststructuralists)—though it should be noted that the debate about women's interests in stem cell research is by no means limited to those who regard themselves as feminists or who can be said to propagate a consistently feminist agenda. That said, the structuralist tradition in feminist scholarship, characterized as it is by a "rejection of masculinist technologization," has been joined by robust poststructuralist engagements, characterized by the "recasting of agency," wherein women are viewed as not simply acted upon but as acting within social worlds. But unlike scholarship and advocacy directed at assisted reproductive technologies, where feminist interventions are predominantly characterized by a reciprocal agenda that seeks to maximize women's agency, an agenda that seeks primarily to protect women from harms has dominated the feminist framing of stem cell research. See Thompson, C. 2002. "Fertile Ground: Feminists Theorize Fertility." In *Infertility Around the Globe: New Thinking on Childlessness, Gender, and Reproductive Technologies*, edited by M. C. Inhorn and F. V. Balen. Berkeley: University of California Press, pp. 53–54.

34. See Bok, H., Schill, K., and Faden, R. 2004. "Justice, Ethnicity, and Stem-cell Banks." *The Lancet 364* (9429): 118–21; Faden, R. R., Dawson, L., Bateman-House, A. S., et al. 2003. "Public Stem Cell Banks: Considerations of Justice in Stem Cell Research and Therapy." *Hastings Center Report 33* (6): 13–27.

35. Rothman, B. K. 2006. "Stem Cell Research: Rethinking the Questions." *Bioethical Inquiry 3*: 16.

36. Interview. Nov. 10, 2005. UC Berkeley Bancroft Library, Prop. 71 Oral History Project.

37. Thompson 2007, p. 207.

38. *Ibid.*, p. 208.

39. Romney, L. 2006. "New Battle Lines Are Drawn over Egg Donation." *Los Angeles Times*, Sept. 13.

40. Goodwin, M. 2007. "Private Ordering and Intimate Spaces: Why the Ability to Negotiate Is Non-Negotiable." *Michigan Law Review 5*: 3.

41. *Ibid.*, p. 4.

42. Thompson 2007, p. 208.

43. In *Stem Cell Century* (Yale University Press, 2007) legal scholar Russell Korobkin comments that "[t]he flawed, magical thinking that underlies this reasoning should be obvious. Wishing away difficult or unpleasant choices in *no* way assists the people who face the choices" (187).

44. See Hayden, C. 2003. *When Nature Goes Public: The Making and Unmaking of Bioprospecting in Mexico.* Princeton, NJ: Princeton University Press.

45. Jan. 30, 2006. CIRM Standards Working Group Meeting. Luxe Hotel, Los Angeles. Field note transcription 1 at: http://www.cirm.ca.gov/files/transcripts /pdf/2006/01-30-06.pdf

46. Jan. 30, 2006. CIRM Standards Working Group Meeting. Luxe Hotel, Los Angeles. Field note transcription at: http://www.cirm.ca.gov/files/transcripts /pdf/2006/01-30-06.pdf

47. *Ibid.*, pp. 174–76.

48. California Prop. 209 (1996) prohibits public institutions from "discriminating" or "granting preferential treatment" on the basis of race, sex, national origin, color, or ethnicity, which in practice leads public institutions to be very cautious in conducting outreach to minority populations.

49. Adding to the unstable political climate toward stem cell research in the United States, state science-making is infused in a "politics of inclusion" (Epstein 2007) that demands that biomedical research be tested upon a demographically representative sample. On the federal level, the National Institutes of Health (NIH) Revitalization Act is premised on the notion that particular populations were being unjustly left out of human subject pools. In the early 1990s, women's health advocates criticized the overrepresentation of white men as the "standard body" in clinical trials, and argued that important biological differences were being ignored as long as white men were standing in for the entire range of people that could benefit from new medical therapies. Epstein, S. 2007. *Inclusion: The politics of Difference in Medical Research.* Chicago: University of Chicago Press.

50. Jan. 30, 2006. CIRM Standards Working Group Meeting. Luxe Hotel, Los Angeles. Field note transcription at: http://www.cirm.ca.gov/files/transcripts /pdf/2006/01-30-06.pdf

51. May 3, 2006. CIRM Standards Working Group Meeting. Luxe Hotel, Los Angeles. Field note transcription at: http://www.cirm.ca.gov/files/transcripts /pdf/2006/05-03-06.pdf

52. See Frow 1997, quoted in Note 28, above.

53. Waldby and Mitchell 2006, p. 41.

54. Jan. 30, 2006. CIRM Standards Working Group Meeting. Luxe Hotel, Los Angeles. Field note transcription 1 at: http://www.cirm.ca.gov/files/transcripts /pdf/2006/01-30-06.pdf

55. "We calculate that a stem-cell bank containing homozygous lines with

the 25 most common haplotypes in the USA would provide matches for about 40% of white Americans, but only 7–8% of African-Americans and 3–6% of Asian-Americans. These differences would probably be even greater in European countries, since they have proportionally smaller non-white populations" (Bok, Schill, and Faden 2004, 119). See also a working paper on "Human Tissue and Blood or Organ Donation, Transplantation and Minority Ethnic Communities" (2004, European Science Foundation International Exploratory Workshop, Leicester, UK, Mary Seacole Research Centre) by Nicky Hudson and Mark Johnson, which reports that "[a] survey of 64 licensed treatment centres in the UK offering either sperm donation or egg donation or both, carried out by Murray & Golombok (1999), suggests that there is a shortage of gamete donors from minority ethnic communities, in particular from the 'Asian' groups and in particular a shortage of egg donors" (6).

56. Goodwin, M. 2006. *Black Markets: The Supply and Demand of Body Parts.* Cambridge, UK: Cambridge University Press.

57. May 3, 2006. CIRM Standards Working Group Meeting. Luxe Hotel, Los Angeles. Field note transcription at: http://www.cirm.ca.gov/files/transcripts /pdf/2006/05-03-06.pdf

58. *Ibid.*, pp. 258–59.

59. May 9, 2007. CIRM Standards Working Group Meeting. Miyako Hotel, San Francisco. Field note transcription at: http://www.cirm.ca.gov/files/ transcripts/pdf/2007/05-09-07.pdf

60. Epstein 2007.

61. Fullwiley, D. 2008. "The Biologistical Construction of Race: 'Admixture' Technology and the New Genetic Medicine." *Social Studies of Science 38* (5): 695–735; Fujimura, J. H., Duster, T., and Rajagopalan, R. 2008. "Race, Genetics, and Disease: Questions of Evidence, Matters of Consequence." *Social Studies of Science 38*: 643–56; Montoya, M. 2011. *Making the Mexican Diabetic: Race, Science, and the Genetics of Inequality.* Berkeley: University of California Press; Soo Jin Lee, S. 2005. "Racializing Drug Design: Implications of Pharmacogenomics for Health Disparities." *American Journal of Public Health 95* (12): 2133–38.

62. The assumption that he can indeed make such a determination is the cornerstone of what some scholars are calling "racial profiling" in medical treatment (Lee, Mountain, and Koenig 2001; Schwartz 2001). For many physicians, this alarmist rendering of the uses of race is more accurately characterized as an everyday heuristic that is all the more necessary in time-constrained practices (Davis 2004). Lee, Sandra Soo-Jin, Mountain, Joanna, and Koenig, Barbara. 2001. "The Meanings of Race in the New Genomics: Implications for Health Disparities Research." *Yale Journal of Health Policy, Law and Ethics* (Spring 1): 33–75; Schwartz, R. S. 2001. "Racial Profiling in Medical Research." *New England Journal of Medicine 344*: 1392–93; Davis, M. M. 2004. "Race-based Im-

munization Recommendations and the Potential to Reduce Racial Disparities." *Journal of the American Medical Association 291*: 2253–55.

63. May 9, 2007. CIRM Standards Working Group Meeting. Miyako Hotel, San Francisco. Field note transcription at: http://www.cirm.ca.gov/files/transcripts/pdf/2007/05-09-07.pdf

64. *Ibid.*, p. 123.

65. May 10, 2007. CIRM Standards Working Group Meeting. Miyako Hotel, San Francisco. Field note transcription at: http://www.cirm.ca.gov/files/transcripts/pdf/2007/05-10-07.pdf

66. Thompson, C. 2007. "Why We Should, in Fact, Pay for Egg Donation." *Regenerative Medicine 2* (2): 203–9.

67. Holm, S. 2006. "Who Should Control the Use of Human Embryonic Stem Cell Lines: A Defense of the Donors' Ability to Control." *Bioethical Inquiry 3*: 55–68, p. 60.

68. Waldby, C., and Cooper, M. 2006. *The Biopolitics of Reproduction: Post-Fordist Biotechnology and Women's Clinical Labor.* Working Paper No.15, Oct. Global Biopolitics Research Group.

69. Stein, Rob. "New York to Pay Women Who Give Eggs to Stem Cell Research." *Washington Post*, June 26 2009. Available at: http://www.washingtonpost.com/wp-dyn/content/article/2009/06/25/AR2009062501931.html

70. In "Paying Women for Their Eggs for Use in Stem Cell Research" (2010), Pamela Foohey explains that "on October 9, 2009, Feminists Choosing Life of New York ('FCLNY'), a self-described 'pro-life feminist' organization, filed suit in New York State court to block the use of state funds to pay women who supply their eggs for stem cell research. FCLNY argues that the compensation program 'provides significant monetary inducements to women to engage in [a] painful and risky procedure, which in part disproportionately appeals to economically vulnerable women,' while 'fail[ing] to satisfactorily provide for informed consent and other safeguards to ensure adequate disclosure to women of the risks of egg harvesting.'" (902–3)

71. http://stemcell.ny.gov/sites/default/files/documents/files/ESSCB_Statement_on_Compensation_of_Oocyte_Donors.pdf

72. *Ibid.*

73. *Ibid.* "When women donate their oocytes for reproductive purposes (i.e., for *in vitro* fertilization), New York State permits reasonable reimbursements for out-of-pocket expenses, time, burden and discomfort associated with the donation, in amounts consistent with the guidelines developed by the American Society for Reproductive Medicine (ASRM). Such reimbursements are widely accepted as ethical, so long as they are not made contingent upon the quality or number of oocytes retrieved, the amount does not act as an undue inducement to donate, and the short- and long-term risks and benefits of donation are fully disclosed to the donor. There is no principled reason to distinguish

between donation of oocytes for reproductive purposes and research purposes when determining the ethicality of reimbursement. The risks associated with donating oocytes to stem cell research are no greater than those associated with reproductive donations. Moreover, donating oocytes to stem cell research arguably confers a greater benefit to society than does oocyte donation for private reproductive use." (1–2)

74. Here I apply Duster's "backdoor" metaphor to indicate the unintended ways in which social division and stratification may be fueled by the (often well-meaning) pursuits of scientists, advocates, and policy actors. See Duster, T. 2003. *Backdoor to Eugenics*. New York and London: Routledge.

75. It is worth noting that while estimates put the proportion of IVF embryos that are derived using paid gametes at only about 15 percent of the "left-over" total, researchers have reason to think that the quality of this 15 percent is generally superior to the 85 percent derived from the couples seeking IVF treatment, because the gametes in the former group are donated from individuals who are generally younger and healthier.

76. Given the institutional culture of the NIH, it is no surprise that the California stem cell agency's first president, a former NIH director, was outspoken in his commitment to ensuring a diverse donor pool. So while the non-compensation agenda clearly resulted in strict prohibitions against coercing socially subordinate women to become donors, he was instrumental in including a diversity clause within the grant process, in the hope that when stem cell therapies reached the clinic, they would be appropriate for a genetically diverse population. The agency's current president, by contrast, is considered by many as the "father of IVF," so it is no surprise that he led the charge to lower the private-public firewall soon after he took the position. The change in leadership corresponds to a shift away from public science-making and the "politics of inclusion" toward a more porous engagement with the IVF industry as a ready resource for stem cell research.

Chapter Four

Portions of this chapter were previously published in Ruha Benjamin, 2011, "Organized Ambivalence: When Stem Cell Research and Sickle Cell Disease Converge," *Ethnicity and Health 16* (4–5): 447–63. Reprinted by permission of the publisher.

1. Pseudonym. For original quotation, see Levine, C. 1996. "Changing Views of Justice After Belmont: AIDS and the Inclusion of 'Vulnerable' Subjects." In *The Ethics of Research Involving Human Subjects: Facing the 21st Century*, edited by H. Y. Vanderpool. Hagerstown, MD: University Publishing Group, pp. 105–26.

2. Field note entry. Aug. 26, 2006. California Institute for Regenerative Medicine (CIRM) Diversity Focus Group.

3. At the time of my fieldwork, the 2003 National Healthcare Disparities Report indicated that "[f]or African Americans, quality of care was poorer than that for whites for 20 out of 46 measures (43%), while care was better quality than whites for just 5 out of 46 measures (11%). . . . Asians and Pacific Islanders had better quality than whites for 12 of the 32 available measures (38%) but still had poorer quality for 7 out of 32 measures (22%)" (38). For a summary of the report, see Kaiser Family Foundation. 2007. "Key Facts: Race, Ethnicity, and Medical Care," pp. 37–40. Available at: http://www.kff.org/minorityhealth/upload/6069-02.pdf

4. A pseudonym; hereafter "Garvey."

5. Bourdieu, P. 1989. "Social Space and Symbolic Power." *Sociological Theory 7* (1): 14–25.

6. Field note entry. Jan. 11, 2006.

7. Rabow, J., Berkman, S. L., and Kessler, R. 1983. "The Culture of Poverty and Learned Helplessness: A Social Psychological Perspective." *Sociological Inquiry 53* (4): 419–34.

8. Powe, B. D., and Johnson, A. 1995. "Fatalism as a Barrier to Cancer Screening Among African-Americans: Philosophical Perspectives." *Journal of Health and Religion 34* (2): 119–26.

9. Dula, A. 1994. "African American Suspicion of the Healthcare System Is Justified—What Do We Do About It?" *Cambridge Quarterly of Healthcare Ethics 3*: 347–57; Gamble, V. 1993. "A Legacy of Distrust: African-Americans and Medical Research." *American Journal of Preventive Medicine 9* (6): 35–38; 1997. "Under the Shadow of Tuskegee: African Americans and Health Care." *American Journal of Public Health 87*: 1773–78.

10. Braunstein, J. B., et al. 2008. "Race, Medical Research Distrust, Perceived Harm and Willingness to Participate in Cardiovascular Prevention Trials." *Medicine 87* (1): 1–9; Corbie-Smith, G., et al. 1999. "Attitudes and Beliefs of African Americans Toward Participation in Medical Research." *Journal of General Internal Medicine 14* (9): 537–46; Corbie-Smith, G., et al. 2002. "Distrust, Race, and Research." *Archives of Internal Medicine 162*: 2458–63.

11. Benjamin, R. 2011. "Organized Ambivalence: When Stem Cell Research and Sickle Cell Disease Converge." *Ethnicity and Health 16* (4–5): 447–63; Hill, S. A. 1994. *Managing Sickle Cell Disease in Low Income Families*. Philadelphia: Temple University Press; Randall, V. R. 1995. "Slavery, Segregation and Racism: Trusting the Health Care System Ain't Always Easy—An African American Perspective on Bioethics." *St. Louis University Public Law Review 15*: 191–236.

12. Illustrative of this trend, Braunstein and colleagues (2008) report that among 717 study participants (36% African American, 64% white), "African American participants more frequently reported that doctors would [not] fully explain research participation to them (24% vs. 13%, p < 0.001), [would] use them as guinea pigs without their consent (72% vs. 49%, p < 0.001), [would] pre-

scribe medication as a way of experimenting on people without their knowledge (35% vs. 16%, p < 0.001), and [would] ask them to participate in research even if it could harm them (24% vs. 15%, p = 0.002). African American participants also more often believed they could [not] freely ask their doctor questions (8% vs. 2%, p < 0.001), and that doctors had previously experimented on them without their consent (58% vs. 25%, p < 0.001)." This study finds that even after controlling for race, sex, socioeconomic status, and disease risk profiles (as individual characteristics and not dynamic processes of objectification and agency), African Americans continue to express less willingness to participate than do white participants.

13. Harris, Y., et al. 1996. "Why African Americans May Not Be Participating in Clinical Trials." *Journal of the National Medical Association 88* (10): 630–34.

14. See Steven Epstein's (2007) concept of "recruitmentology" in *Inclusion: The Politics of Difference in Medical Research*. Chicago: University of Chicago Press.

15. Levine 1996.

16. Jasanoff, S. 2005. "In the Democracies of DNA: Ontological Uncertainty and Political Order in Three States." *New Genetics and Society 24* (2): 139–56; Timmermans, S., and Angell, A. 2001. "Evidence-Based Medicine, Clinical Uncertainty, and Learning to Doctor." *Journal of Health and Social Behavior 42* (4): 342–59.

17. Haraway, D. 1991. "Situated Knowledges." In *Simians, Cyborgs, and Women: The Reinvention of Nature*. New York: Routledge.

18. I acknowledge Anne Pollock (assistant professor of science, technology, and culture, Georgia Tech), whose comments on a much earlier version of this chapter provided me with this insight.

19. For a full account of the role of uncertainty in sickle cell patient care, see Carolyn Rouse. 2009. *Uncertain Suffering: Racial Health Disparities and Sickle Cell Disease*. Berkeley: University of California Press.

20. See U.S. Food and Drug Administration. 2005. "Postmarket Drug Safety Information for Patients and Providers." Available at: http://www.fda.gov /Drugs/DrugSafety/PostmarketDrugSafetyInformationforPatientsandProviders/ ucm108111.htm

21. Personal interview. Dec. 7, 2005.

22. Whitmarsh, I. 2008. "Biomedical Ambivalence: Asthma Diagnosis, the Pharmaceutical, and Other Contradictions in Barbados." *American Ethnologist 35* (1): 49–63.

23. *Ibid.*, p. 58.

24. Hill, S. A. 1994. *Managing Sickle Cell Disease in Low-Income Families*. Philadelphia: Temple University Press, p. 107.

25. Collins, F. S., and Guttmacher, A. E. 2007. "Foreword." In *Renaissance of Sickle Cell Disease Research in the Genome Era*, edited by Betty Pace. London: Imperial College Press, pp. xxix–xxx.

26. Personal interview. Dec. 7, 2005.

27. Levin, J., et al. 2005. "Religion, Health and Medicine in African Americans: Implications for Physicians." *Journal of the National Medical Association* *97* (2): 237–49.

28. Harrison, M. O., et al. 2005. "Religiosity/Spirituality and Pain in Patients with Sickle Cell Disease." *Journal of Nervous and Mental Disease 193* (4): 250–57.

29. Hill, S. A. 1994. *Managing Sickle Cell Disease in Low-Income Families.* Philadelphia: Temple University Press, p. 98.

30. For a fuller discussion of the role of school climate in sickle cell health outcomes, see Dyson, S., et al. 2007. "The Educational Experiences of Young People with Sickle Cell Disorder: A Commentary on the Existing Literature." *Disability and Society 22* (6): 581–94.

31. Hill 1994, pp. 97–98.

32. See Charis Thompson's (2005) notion of "strategic naturalization" in *Making Parents: The Ontological Choreography of Reproductive Technologies* (Cambridge, MA: MIT Press).

33. Wade, P. 2002. Race, *Nature and Culture: An Anthropological Perspective.* London: Pluto Press.

34. Hasnain-Wynia, R., et al. 2007. "Disparities in Health Care Are Driven by Where Minority Patients Seek Care." *Archives of Internal Medicine 167*: 1233–39.

35. Interview cited in study report: Duster and Beeson. 1997. "Pathways and Barriers to Genetic Testing: Molecular Genetics Meets the 'High-Risk Family.'" University of California Berkeley Institute for the Study of Social Change. Available at: http://www.ornl.gov/sci/techresources/Human_Genome/resource/duster.html

36. Anionwu, E. N. 1993. "Sickle Cell and Thalassaemia: Community Experiences and Official Response." In *"Race" and Health in Contemporary Britain*, edited by W.I.U. Ahmad. Buckingham, UK: Open University Press, pp. 76–95; Black, J., and Laws, S. 1986. *Living with Sickle Cell Disease.* London: East London Sickle Cell Society; Duster, T. 1990. *Backdoor to Eugenics.* New York and London: Routledge; Hill 1994; Jones, J. H. 1993. *Bad Blood: The Tuskegee Syphilis Experiment.* New York: Free Press.

37. Rouse, C. M. 2004. "Paradigms and Politics: Shaping Health Care Access for Sickle Cell Patients Through Discursive Regimes of Biomedicine." *Culture, Medicine, and Psychiatry 28*: 371.

38. Anionwu, E. N., and Atkin, K. 2001. *The Politics of Sickle Cell and Thalassemia.* Buckingham, UK: Open University Press.

39. From "Fighting the Stigma: Pain in Sickle Cell Disease—Roxanne 3," online interview. Available at: http://www.youtube.com/watch?v=lI2amLqhKec

40. Field note entry. Feb. 10, 2010.

41. Roberts, D. 1998. *Killing the Black Body: Race, Reproduction, and the Meaning of Liberty.* New York: Vintage.

42. For a full historical account of how healthcare and medical research are conflated in the treatment of African American subjects, see Washington, H. A. 2006. *Medial Apartheid: The Dark History of Medical Experimentation on Black Americans from Colonial Times to the Present.* New York: Doubleday.

43. Field note entry. Dec. 5, 2005.

44. Waldby, C., and Mitchell, R. 2006. *Tissue Economies: Blood, Organs, and Cell Lines in Late Capitalism.* Durham, NC: Duke University Press.

45. Faden, R. R., et al. 2003. "Public Stem Cell Banks: Considerations of Justice in Stem Cell Research and Therapy." *Hastings Center Report 33* (6): 13–27.

46. Wailoo, K. 2001. *Dying in the City of Blues: Sickle Cell Disease.* Chapel Hill: University of North Carolina Press, p. 11.

47. *Ibid.*, p. 8.

48. "Marrow Transplant Found to Be a Cure in Sickle Cell Case," Sept. 20, 1984; cited in Wailoo, K., and Pemberton, S. 2006. *The Troubled Dream of Genetic Medicine: Ethnicity and Innovation in Tay-Sachs, Cystic Fibrosis, and Sickle Cell Disease.* Baltimore: Johns Hopkins University Press, p. 148.

49. Wailoo and Pemberton (2006), p. 149.

50. See Fortun, M. 2008. *Promising Genomics: Iceland and deCODE Genetics in a World of Speculation.* Berkeley: University of California Press.

51. Hall, C. T. 2005. "Stem Cells: The $3 Billion Bet; One Man's Scientific Mission: Housing Developer Leads California's Research Effort." *San Francisco Chronicle.* Available at: http://articles.sfgate.com/2005-04-11/news/17367162_1_stem-cell-cell-research-cell-program

52. Personal interview with H. Rex Greene conducted by Marc Strassman. Oct. 21, 2004. "California Politics Today #153." Audio link available at: http://etopiamedia.net/empnn/pages/cpt-emnn/cpt-emnn153-5551212.html

53. See, for example, Beeson, D., and Lipman, A. 2006. "Egg Harvesting for Stem Cell Research: Medical Risks and Ethical Problems." *Reproductive BioMedicine Online 13* (4): 573–79.

54. The board composition includes ten patient advocate seats, one each for Type II Diabetes, Alzheimer's, Cancer, Type I Diabetes, Parkinson's Disease, MS/ALS, HIV/AIDS, Mental Health, Spinal Cord Injury, and Heart Disease.

55. Field note entry. Jan. 15, 2008. Excerpt from correspondence that took place between Greenlining Institute and the stem cell agency's grant review committee.

56. Brown, M. B., and Guston, D. H. 2005. "Science, Democracy, and the Right to Research." *Science and Engineering Ethics 15* (3): 1471–1546.

57. Chapter 6 will elaborate on this model of "power-sharing" put forth in Winickoff, D. E. 2008. "From Benefit Sharing to Power Sharing: Partnership Governance in Population Genomics Research." UC Berkeley, Center for the Study of Law and Society, Jurisprudence and Social Policy Program, p. 1. Available at: http://escholarship.org/uc/item/845393hh

58. Field note. Feb. 6, 2008. "Institutional Landscapes in Stem Cell Research" workshop at UCSF Mission Bay Conference Center. The workshop was sponsored by the Stem Cell Center and the Science, Technology, and Society Center, both at UC Berkeley.

59. Good, M-J. D. 2001. "The Biotechnical Embrace." *Culture, Medicine and Psychiatry 25*: 395–410.

60. Whitmarsh 2008.

Chapter Five

1. Johnson, J. M., and Melnikov, A. 2009. "The Wisdom of Distrust." In *Studies of Symbolic Interaction*, vol. 33, edited by N. K. Denzin. Bingley, UK: Emerald Group.

2. See Brownlie, J., Greene, A., and Howson, A. (eds.). 2008. "Introduction." In *Researching Trust and Health*. New York: Routledge.

3. "The use of deadly force has been at the heart of tensions between police and Oakland's communities of color for decades." See Winston, Ali. 2011. "Deadly Secrets: How California Law Shields Oakland Police Violence." Available at: http://colorlines.com/archives/2011/08/deadly_secrets_how_california_law_has_shielded_oakland_police_violence.html

4. Moore, L. 2008. "Brick by Brick." *Poor Magazine*. Available at: http://www.poormagazine.org/node/2056

5. Alexander, M. 2010. *New Jim Crow: Mass Incarceration in the Age of Colorblindness*. New York: New Press; Wacquant, L. 2009. *Deadly Symbiosis: Race and the Rise of the Penal State*. Cambridge, UK: Polity Press; 2009. *Punishing the Poor: The Neoliberal Government of Social Insecurity*. Durham, NC: Duke University Press.

6. Leroy Moore and a former NYPD police officer who is disabled, Emmitt Thrower, are producing a documentary called "Broken Bodies: Police Brutality Profiling" (Wabi Sabi Productions).

7. Durant, R. W., Legedza, A. T., Marcantonio, E. R., Freeman, M. B., and Landon, B. E. 2011. "Different Types of Distrust in Clinical Research Among Whites and African Americans." *Journal of the National Medical Association 103* (2): 123–30.

8. *Ibid.*

9. For elaboration, see Fraser, N. 1997. Justice Interruptus: Critical Reflections on the "Postsocialist" Condition. New York: Routledge.

10. Wacquant, L. 2008. "Relocating Gentrification: The Working Class, Science and the State in Recent Urban Research." *International Journal of Urban and Regional Research 32* (1): 198–205.

11. Johnson and Melnikov 2009, p. 16.

12. Wasserman, J., Flannery, M. A., and Clair, J. M. 2007. "Raising the

Ivory Tower: The Production of Knowledge and Distrust of Medicine Among African Americans." *Journal of Medical Ethics 33* (3): 177–80.

13. Royal, C.D.M., et al. 2000. "Recruitment Experience in the First Phase of the African American Hereditary Prostate Cancer (AAHPC) Study." *Annals of Epidemiology 10*: S68–S77.

14. Swanson, G. M., and Ward, A. J. 1995. "Recruiting Minorities into Clinical Trials: Toward a Participant-Friendly System." *Journal of the National Cancer Institute 87* (23): 1747–59.

15. Gorelick, P. B., Harris, Y., Burnett, B., and Bonecutter, F. J. 1998. "The Recruitment Triangle: Reasons Why African Americans Enroll, Refuse to Enroll, or Voluntarily Withdraw from a Clinical Trial." *Journal of the National Medical Association 90* (3): 141–45.

16. Feb. 3, 2005. Regular ICOC Meeting. Neurosciences Institute, San Diego. Available at: http://www.cirm.ca.gov/files/transcripts/pdf/2005/02-03-05.pdf

17. See U.S. Department of Health and Human Services. Office of Extramural Research. National Institutes of Health. 2001. "NIH Policy and Guidelines on the Inclusion of Women and Minorities as Subjects in Clinical Research—Amended" (Oct.). Available at: http://grants.nih.gov/grants/funding/women_min/guidelines_amended_10_2001.htm

18. From "CIRM Grants Administration Policy for Academic and Non-Profit Institutions": "Since a primary aim of research is to provide scientific evidence leading to a change in health policy or standard of care, it is imperative to determine whether the intervention or therapy being studied affects women or men or members of minority groups and their subpopulations differently. This requirement ensures that all CIRM-funded clinical research will be carried out in a manner sufficient to elicit information about individuals of both sexes/genders and diverse racial and ethnic groups and, particularly in clinical trials, to examine differential effects on such groups." Available at: http://www.cirm.ca.gov/reg/pdf/reg100500_policy.pdf

19. See Braun, Lundy, Fausto-Sterling, Anne, Fullwiley, Duana, Hammonds, Evelynn M., Nelson, Alondra, Quivers, William, Reverby, Susan M., and Shields, Alexandra E. 2007. "Racial Categories in Medical Practice: How Useful Are They?" *PLoS Med 4* (9). See also Epstein, Steven. 2007. *Inclusion: The Politics of Difference in Medical Research.* Chicago: University of Chicago Press; Koenig, B. A., Soo-Jin Lee, S., and Richardson, S. S., eds. 2008. *Revisiting Race in a Genomic Age.* Piscataway, NJ: Rutgers University Press; Krimksy, S., and Sloan, K., eds. 2001. *Race and the Genetic Revolution: Science, Myth, and Culture.* New York: Columbia University Press; Roberts, Dorothy. 2008. "Is Race-based Medicine Good for Us? African American Approaches to Race, Biomedicine, and Equality." *Journal of Law, Medicine, and Ethics 36*: 537–45; Shim, J. K. 2005. "Constructing 'Race' Across the Science-Lay Divide: Racial Formation in the Epidemiology and Experience of Cardiovascular Disease." *Social*

Studies of Science 35 (3): 405–36; Whitmarsh, I., and Jones, D., eds. 2010. *What's the Use of Race? Modern Governance and the Biology of Difference.* Cambridge, MA: MIT Press.

20. http://www.cirm.ca.gov/files/MeetingReports/Diversity_Workshop_Report_4_23_10b.pdf

21. *Ibid.*

22. Moller, Mark S. 2008. "Human Embryonic Stem Cell Research, Justice, and the Problem of Unequal Biological Access." *Philosophy, Ethics, and Humanities in Medicine 3*: 22.

23. Satel, S. 2002. "I Am a Racially Profiling Doctor." *New York Times* (May 5). Available at: http://www.nytimes.com/2002/05/05/magazine/i-am-a-racially-profiling-doctor.html?pagewanted=all&src=pm

24. Epstein, S. 2007. "To Profile or Not to Profile: What Difference Does Race Make?" In Epstein 2007, pp. 203–32.

25. Wasserman, Flannery, and Clair 2007.

26. Feb. 26, 2010. CIRM Diversity Workshop, Charles Drew University of Medicine and Science, Los Angeles. Field note transcription at: http://www.cirm.ca.gov/files/MeetingReports/Diversity_Workshop_Report_4_23_10b.pdf

27. Durant, Legedza, Marcantonio, Freeman, and Landon 2011.

28. Royal et al. 2000.

29. Field note. Oct. 31, 2005.

30. Field note entry. Oct. 31, 2005. Notably, Sethe Hart (see Ch. 4) mentions rejecting this elective spleen surgery for Destiny (Dec. 7, 2005).

31. Field note entry. Nov. 2, 2004. Garvey sickle cell clinic.

32. Lathrop, Douglas. 2010. "The Education of Professir X" (Feb.). Available at: http://www.newmobility.com/articleView.cfm?id=11582

33. *Ibid.*

34. Chadwick, John. 2010. "Neuroscientist Wise Young and Hip-Hop Artist Richard Gaskin Find Common Cause as Spinal Cord Injury Activists" (July 20). Available at: http://www.newjerseynewsroom.com/healthquest/neuroscientist-wise-young-and-hip-hop-artist-richard-gaskin-find-common-cause-as-spinal-cord-injury-activists

35. *Ibid.*

36. Alcoff, L. 1991. "The Problem of Speaking for Others." *Cultural Critique 20*: 5–32.

37. Saward, M. 2006. *The Representative Claim.* Oxford University Press., p. 298.

38. Callon, M. 2005. Domestication of the Scallops and the Fishermen of St Brieuc Bay." In *Knowledge: Critical Concepts*, edited by Nico Stehr and Reiner Grundmann. Routledge. pp. 14–15.

39. Saward, M. 2006. *The Representative Claim.* Oxford, UK: Oxford University Press, pp. 15, 17.

40. Bourdieu, P. 1991. "The Peculiar History of Scientific Reason." *Sociological Forum 6* (1): 3–26.

41. Lee, O. 1998. "Culture and Democratic Theory: Toward a Theory of Symbolic Democracy." *Constellations 5* (4): 433–55, 434.

42. According to Paolo Mancini and David Swanson (1999), the "breakdown of traditional social structures under the strains of modernization has created a need for new forms of political communication in which new 'symbolic realties' have to be created containing symbolic templates of heroes and villains, honored values and aspirations, histories, mythologies, and self-definition." Cited in Street, J. 2004. "Celebrity Politicians: Popular Culture and Political Representation." *British Journal of Politics and International Relations 6*: 435–52, 441.

43. Sheila Jasanoff notes, in particular, that this competent subject "struggles with ambivalence in the face of competing cognitive and social pressures, robustly copes with ignorance and uncertainty as well as "the facts," and reserves the right to make moral choices about the purposes and governance of technology. . . . [This is a] culturally knowledgeable figure, able to master a more complex phenomenology in some respects than that of science; in particular, lay citizens may be better than experts at making room for the unknown along with the known." (254) See Jasanoff, S. 2005. *Designs on Nature: Science and Democracy in Europe and the United States*. Princeton, NJ: Princeton University Press.

44. Sociologist Barbara Misztal explains that "Trust is usually defined as confidence that partners will not exploit each other's vulnerability. While stressing that vulnerability cannot be conceived as a single continuum," she elaborates how "vulnerability [is] irreducibly plural and rooted in the human condition of dependence on others, in the unpredictability of action and in the irreversibility of human experiences." See Misztal, B. A. 2011. "Trust: Acceptance of, Precaution Against and Cause of Vulnerability." *Comparative Sociology 10* (3): 358–79.

45. Rapp, R. 1999. *Testing Women, Testing the Fetus: The Social Impact of Amniocentesis in America*. New York: Routledge, p. 306.

46. *Ibid.*, p. 309.

47. Kittay 1997, cited in Rapp 1999, p. 308.

48. Marx, K. 1852. *The Eighteenth Brumaire of Louis Bonaparte*. Available at: http://www.marxists.org/archive/marx/works/1852/18th-brumaire/index.htm

49. Du Bois, W.E.B. 1903. *The Souls of Black Folk*. Lindenhurst, NY: Tribeca Books.

50. Tervalon, M., and Murray-Garcia, J. 1998. "Cultural Humility Versus Cultural Competence: A Critical Distinction in Defining Physician Training." *Journal of Health Care for the Poor and Underserved 9* (2): 117–24, 117.

51. California Health Advocates. 2007. "Are You Practicing Cultural Humility?—The Key to Success in Cultural Competence" (Apr.). Available at: http://www.cahealthadvocates.org/news/disparities/2007/are-you.html

52. Jasanoff, S. 2007. "Technologies of Humility." *Nature 450* (1): 33.

Chapter Six

1. Winner, L. 1978. *Autonomous Technology: Technics-out-of-Control as a Theme in Political Thought.* Cambridge, MA: MIT Press, p. 314.

2. King is credited with saying these famous words in a Poor People's Campaign speech organized by the Southern Christian Leadership Conference in 1968.

3. Phlippidis, A. 2008. "Calif. State Senator Vows to Push Life-Sci Bill Without More Stem-Cell Amendments." *BioRegion News.* Available at: http://www.geneticsandsociety.org/article.php?id=4198

4. Newmark, C. 2004. "Good Info on California Prop. 72." Available at: http://craigconnects.org/2004/10/good_info_on_ca.html

5. Available at: http://www.followthemoney.org/database/StateGlance/committee.phtml?c=1180

6. Available at: http://www.followthemoney.org/database/StateGlance/committee.phtml?c=1224

7. Available at: http://goliath.ecnext.com/coms2/gi_0199-2329737/No-vote-Urged-on-California.html

8. Birman, I. 2004. "Proposition 72: A Misadventure We Can Ill Afford." Available at: http://www.freerepublic.com/focus/f-news/1249058/posts

9. "'No' Vote Urged on California Prop. 72 on November 2." *Franchising World.* Available at: http://goliath.ecnext.com/coms2/gi_0199-2329737/No-vote-Urged-on-California.html

10. Goodrich, B. 2004. "Proposition 72: Agribusiness Takes Closer Look at Health Insurance Act." Available at: http://westernfarmpress.com/proposition-72-agribusiness-takes-closer-look-health-insurance-act

11. The group opposing SB 2 raised $9.9 million "to launch a toughly worded TV ad campaign"—"considerably more" than the $2.1 million raised by those in favor of the measure, according to the *Hartford Courant.* SB 2 would "require employers with 200 or more employees to provide health insurance to workers and their dependents by 2006 or pay into the state fund. Employers with 50 to 199 employees [would] have to provide health insurance only to workers by 2007. Companies with fewer than 20 workers [would] not have to comply with the law, and the law also [would] exempt employers with 20 to 49 workers unless the state provide[d] them with tax credits to offset the cost of health coverage." Those opposed to SB 2 also "appear to have history on their side," according to the *Courant*, as groups in the past have been successful in defeating similar requirements." Available at: http://www.californiahealthline.org/articles/2004/9/27/proposition-72-campaigns-examined.aspx#ixzz1jGzKowy2; for a brief description of No and Yes on 72 ads, see Gledhill, Lynda. 2004. "AdWatch: Proposition 72." Available at: http://articles.sfgate.com/2004-10-05/bay-area/17448190_1_health-care-government-run-health-plan-government-run-health

12. "October 2004 Action Alert—Proposition 72." Available at: http://

www.cabwhp.org/resources/issue_guides/october-2004-action-alert-proposi
tion-72

13. Lusvardi, W. 2011. "Cancel Prop. 71 Stem Cell Funding?" Available at:
http://www.calwatchdog.com/2011/02/23/cancel-prop-71s-stem-cell-research
-funding/

14. The purpose of Ortiz's SB 18 was to institute a routine performance au-
dit to evaluate the implementation of Prop. 71. The primary purpose of Ortiz's
SB 1260, "Reproductive Health and Research," was to create protections for
research subjects, prohibit payment for human oocytes, require physicians to
provide donors with a written statement of the health impacts associated with
donation, require written consent by donors, require stricter research oversight
procedures and annual updates, and give more control to the state Department of Human Services than was originally written into the text of Prop. 71.
(This bill was passed almost unanimously by the legislature and signed into law
by Governor Schwarzenegger in September 2006.) And the primary purpose
of Ortiz's Senate Constitutional Amendment (SCA) 13 was to impose stricter
guidelines that would prevent conflicts of interest among board members (by
requiring more financial disclosure), guarantee greater adherence to open meeting laws, and ensure that more revenue from future therapies was returned to
the state to benefit low-income Californians. (This bill was never approved,
largely because term limits forced Senator Ortiz from office at the end of 2006,
which also meant that she was losing legislative clout in her final months.)

15. Interview with Paul Berg. Feb. 16, 2006. UC Berkeley Bancroft Library.
Prop. 71 Oral History Project.

16. California Disability Community Action Network. 2006. *Disability Rights News Report*. Available at: http://www.cdcan.us/cdcan_reports
/2006/040-2006.htm

17. Jensen, D. 2006. "Text of the Klein-Ortiz-SB401 Letter." *California Stem Cell Report* (June 10). Available at: http://californiastemcellreport.
blogspot.com/2006/06/text-of-klein-ortiz-sb401-letter.html. For an examination of the Americans for Cures letter, see "Klein the Facts of the Matter." Available at: http://californiastemcellreport.blogspot.com/search?q=SB+401

18. See CIRM transcript. Available at: http://www.cirm.ca.gov/Transcript
_060605

19. June 6, 2005. Regular ICOC Meeting. Sacramento Convention Center,
Sacramento, California. Field note transcription at: http://www.cirm.ca.gov/
files/transcripts/pdf/2005/06-06-05.pdf

20. *Ibid.*, p. 70.

21. *Ibid.*, pp. 100–101.

22. May 23, 2005. Regular ICOC Meeting. The Tech Museum of Innovation, San Jose, California. Field note transcription at: http://www.cirm.ca.gov/
files/transcripts/pdf/2005/05-23-05.pdf

23. June 6, 2006. Comments before the Regular ICOC Meeting. Sacramento Convention Center, Sacramento, California. Available at: http://www .cirm.ca.gov/files/transcripts/pdf/2005/06-06-05.pdf

24. Somers, T. 2006. "Initiative's Creator Keeps Tinkering: Senator, Oversight Panel Often at Odds." Available at: http://www.geneticsandsociety.org/article.php?id=1879

25. June 6, 2005. Regular ICOC Meeting. Sacramento Convention Center, Sacramento, California. Field note transcription at: http://www.cirm.ca.gov/files/transcripts/pdf/2005/06-06-05.pdf

26. Longaker, M. T., Baker, L. C., and Greely, H. T. 2007. "Proposition 71 and CIRM: Assessing the Return on Investment." *Nature Biotechnology 25* (5): 520.

27. Howard, J. 2010. "Stem Cell Agency Draws Capitol Scrutiny–Again." Available at: http://www.geneticsandsociety.org/article.php?id=5122

28. Letter written by Senator Sheila Kuehl, posted on the California Stem Cell Report on Dec. 11, 2007. Available at: http://californiastemcellreport.blog spot.com/2007/12/text-of-lawmakers-letter.html

29. Text of SB 1565, introduced by Senators Kuehl and Runner on Feb. 22, 2008. Available at: http://www.leginfo.ca.gov/pub/07-08/bill/sen/sb_1551-1600/sb_1565_bill_20080222_introduced.html

30. Philippidis, A. 2008. "Calif. State Senator Vows to Push Life-Sci Bill Without More Stem-Cell Amendments." Available at: http://www.geneticsand society.org/article.php?id=4198

31. Howard 2010.

32. Leonhardt, D. 2010. "In Health Care Bill, Obama Attacks Wealth Inequality." *New York Times* (Mar. 23). Available at: http://www.teamsters952.org/In_Health_Bill_Obama_Attacks_Wealth_Inequality_-_NYTimes.c.pdf

33. Moller, M. S. 2008. "Human Embryonic Stem Cell Research, Justice, and the Problem of Unequal Biological Access." *Philosophy, Ethics, and Humanities in Medicine 3*: 22.

34. *Ibid.*

35. Woolf, S. H., Johnson, R. E., Fryer, G. E., Rust, G., and Satcher, D. 2004. "The Health Impact of Resolving Racial Disparities: An Analysis of US Mortality Data." *American Journal of Public Health 94* (12): 2078–81.

36. Interview with January W. Payne, quoted in Payne, J. W. 2004. "Dying for Basic Care: For Blacks, Poor Health Care Access Cost 900,000 Lives." *Washington Post* (Dec. 21).

37. Woolf et al., 2004.

38. Interview with Payne, 2004.

39. Satcher, D., Fryer, G. E., Jr., McCann, J., Troutman, A., Woolf, S. H., and Rust, G. 2005. "What If We Were Equal? A Comparison of the Black-White Mortality Gap in 1960 and 2000." *Health Affairs 24* (2): 459–64.

40. Dahlberg, L. 2005. "The Habermasian Public Sphere: Taking Difference Seriously?" *Theory and Society 34* (2): 111–36.

41. For an elaboration of "outsiders-within," see Hill Collins, P. 1990. *Black Feminist Thought: Knowledge, Consciousness, and the Politics of Empowerment.* New York: Routledge.

42. Lister, R. 2003. *Citizenship: Feminist Perspectives.* New York: NYU Press.

43. Nowotny, H., Scott, P., and Gibbons, M. 2001. *Re-thinking Science: Knowledge and the Public in an Age of Uncertainty.* Cambridge, UK: Polity Press, p. 212.

44. Arbab, F. 2000. "Promoting a Discourse on Science, Religion, and Development." In *The Lab, the Temple, and the Market,* edited by Sharon Harper. Ottawa, Canada: Kumarian Press, pp. 149–238, 153. Available at: http://idl-bnc .idrc.ca/dspace/bitstream/10625/18688/5/116242.pdf

45. *Ibid.,* p. 123.

46. FUNDAEC is a nongovernmental development agency in Cali, Colombia, South America. See http://www.fundaec.org/en/

47. For data on rates of health insurance by employment and race/ethnicity, see "Snapshot: California's Uninsured: 2007." Available at: http://www.chcf .org//media/MEDIA%20LIBRARY%20Files/PDF/S/PDF%20SnapshotUn insured07.pdf

48. American Community Survey (2010) for demographic and geographic breakdown; Current Population Survey Annual Social and Economic Supplement (1970–2011) for trends (both from the U.S. Census Bureau). Census Bureau Supplemental Poverty Measure resources. Available at: http://www.ppic.org /main/publication_show.asp?i=261

49. Lee, H., and McConville, S. 2007. "Death in the Golden State: Why Do Some Californians Live Longer?" *Policy Institute of California 9* (1): 1. Available at: http://www.ppic.org/content/pubs/cacounts/CC_807HLCC.pdf

50. Health Disparities and the Body Politic: A Series of International Symposia. 2005. Available at: http://www.hsph.harvard.edu/disparities/book/ HealthDisparities.pdf

51. Krieger, N. 2009. "Advice to the Next President: 7 Ways to Fight Health Inequities." Available at: http://www.youtube.com/watch?v=apkhG88Z2uI

52. The 2012 annual meeting of the American Sociological Association was entitled "Real Utopias." See http://www.asanet.org/AM2012/meeting_theme.cfm

53. Available at: http://www.ssc.wisc.edu/wright/ERU_files/ERU-CHAP TER-1-final.pdf

54. Talk entitled "Can We Leave the Bauxite in the Mountain? Field Notes on Democracy." Available at: http://www.youtube.com/watch?v=22LNT3H_YjY

55. "Stem Cell Research Focus of Meeting." Available at: http://www.ut san diego.com/news/2008/nov/08/1b8stem213033-stem-cell-research-focus-meet ing/?page=1

56. Jasanoff, S. 2005. *Designs on Natures: Science and Democracy in Europe and the United States*. Princeton, NJ: Princeton University Press.

57. Arbab 2000, p. 206.

58. *Ibid.*, p. 215.

59. *Ibid.*

60. *Ibid.*, p. 171.

61. I would like to acknowledge the work of Zeus Yiamouyiannis at the Interactivity Foundation for expert facilitation of the citizen discussions in which I took part. These were not a part of the California stem cell initiative but were rather a site in which I triangulated my investigation of public participation in science.

62. Emphasis added. Available at: http://www.interactivityfoundation.org/

63. *Ibid.*

64. *Ibid.*

65. Arbab 2000, p. 218.

66. See Bahá'í International Community Statement on the "Elimination Between the Extremes of Poverty and Wealth." Feb. 3, 2012. Available at: http://www.youtube.com/watch?v=IcSKK53T3kU

67. Winickoff, D. E. 2008. "From Benefit Sharing to Power Sharing: Partnership Governance in Population Genomics Research." UC Berkeley, Center for the Study of Law and Society, Jurisprudence and Social Policy Program, p. 1. Available at: http://escholarship.org/uc/item/845393hh

68. John Terborgh, in a review of Tim Flannery's book *Here on Earth: A Natural History of the Planet* (*New York Review of Books*, Oct. 13, 2011). Available at: http://the-scientist.com/2012/01/01/speaking-of-science-10/

69. Arbab 2000, p. 213.

70. *Ibid.*, p. 200.

71. *Ibid.*, p. 218.

72. *Ibid.*, p. 219.

73. Nowotny, Scott, and Gibbons 2011, p. 228.

74. Arbab 2000, p. 213, commenting on Winner's work.

75. Winner 1978, pp. 312–13; emphasis added. "Technology, then, allows us to ignore our own works. It is *license to forget*. In its sphere the truths of all important processes are encased, shut away, and removed from our concern. This more than anything else, I am convinced, is the true source of the colossal passivity in man's dealings with technical means."

76. For an in-depth analysis of the importance of political-cultural recognition and socioeconomic redistribution in democratic life, see Fraser, N. 1997. *Justice Interruptus: Critical Reflections on the "Postsocialist" Condition*. New York: Routledge.

77. Nowotny, Scott, and Gibbons 2011, p. 247. "The lack of attention to power issues in the agora has been brought up in research literature by several

authors (Barré, 2001; Pestre, 2003; Fuller, 2002), but no clear idea of how it can be dealt with has been offered."

78. Gibbons, M. 1999. "Science's New Social Contract with Society." *Nature 402*: C81–C84.

79. Leydesdorff, L., and Etzkowitz, H. 1996. "Emergence of a Triple Helix of University-Government-Industry Relations." Available at: http://www.leydesdorff.net/th1a/

80. Available at: http://socgen.ucla.edu/research/ucla-in-la-grant-for-genetics-and-society-high-school-science-education-outreach/

81. School demographics available at: http://search.lausd.k12.ca.us/cgi-bin/fccgi.exe?w3exec=cbeds3&which=location_code&info=8727

82. Haddow, G., et al. 2007. "Tackling Community Concerns About Commercialisation and Genetic Research: A Modest Interdisciplinary Proposal." *Social Science and Medicine 64*: 272. Cited in Winickoff 2008, p. 10.

83. *Ibid.*, p. 20.

84. *Ibid.*, p. 4.

Acknowledgments

1. Baha'u'llah. 1976. *A Compilation on Bahá'í Education*, compiled by the Research Department of the Universal House of Justice, Haifa, Israel, p. 59.

2. *The Wiz* (dir. Sidney Lumet). 1978. Universal Studios.

INDEX

ableism, 58

able-ist paradigm of stem cell advocates, 64, 71

access of poor to stem cell therapies: bioconstitutionalism and, 31; CIRM lip service to, 40–42; efforts to ensure, 31, 37, 39, 164; elite views on, 31, 37; as issue, 28–30; necessity of, to avoid increases in health disparities, 39; state bill 1565 and, 37–38, 39

access to biomedical goods. *See* right to consume biomedical goods

adult stem cells, as non-controversial, 113

African American dispossession, and distrust of clinical research, 137–38

African American distrust of clinical research: causes of, 137–38; and CIRM re-funding, 140; community-based efforts to cultivate trust, inadequacy of, 139–40; and difficulty of creating diverse cell lines, 142, 144; and difficulty of recruiting clinical research subjects, 140–41; interpersonal *vs.* societal, 136–37; ongoing social production of, 146; as rational response to social conditions, 138–

41, 144–47, 152–53, 155; research on, as misguided, 139. *See also* sickle cell stem cell treatments, low African American participation rates; *entries under* minority distrust

African American distrust of government: government spending priorities and, 136; white perception of, 145–47

African American distrust of health care system: ambivalence, causes of, 117–18, 123–24, 133–34; distrust of medical professionals, 116, 213n12; home remedies, faith in, 119, 121–24; lack of health insurance and, 115; and low utilization of stem cell therapies, 115–18; medications, concerns about, 119–20; perception of modern medicine as fundamentally misguided, 122; as rational response to social conditions, 138–41, 144–47, 152–53, 155. *See also* sickle cell stem cell treatments, low African American participation rates; *entries under* minority distrust

African Americans: annual deaths related to inadequate health care,

amendment and, 15, 18–19, 21; scientists' views on, 4, 178; state initiatives on stem cell research and, 4–5, 12; steps in construction of, 19–20; U.S. lag in developing mechanisms of, 10–11

poverty, racialization of in U.S.: and coercive nature of egg donation, 89; and stigmatization of poor, 137

poverty rate, among California minorities, 170

Preciado, Phyllis, 47–50, 51, 140

preimplantation genetic diagnosis, disabled activists' perception of as eugenics, 67

prisoners, abuse of in scientific research, 3

privatization of public goods: CIRM creation as, 14–15; tax breaks for biotechnology industry and, 14, 190n36

Pro-Choice Alliance for Responsible Research, 81, 84, 89–90, 96, 99

Professir X. See Gaskin, Richard

progress, scientific, exclusion of *demos* and, 16

Proposition 71: and biomedical consumers as default construction of "the public," 179; bond sales authorized by, 5; defining of past and future states in, 23; and failure of Proposition 72, 13, 15, 160; feminist views on, 84, 85–86; goals of, 12; intellectual property provisions, 163; lack of social justice provisions in, 14–15; and mandate claimed by supporters, 172–73; marketing campaign for, 44–45, 46–47, 78; noncompensation policy for egg donors in, 85, 89; passage of, 5, 12; political spaces opened by, 193n10; power

sharing by minority communities, lack of provisions for, 132; sickle cell cure as marketing point for, 129–31. *See also* California Stem Cell Research and Cures Initiative

Proposition 72: campaign against, donors to, 158, 221n11; campaign for, donors to, 158, 221n11; defeat of, 13, 15, 160; importance of, 158; Maddow radio program on, 157; and populist *vs.* elite rights, 160; potential impact of, 158–59, 160; provisions of, 157, 159, 221n11

public, "right," CIRM limiting of stem cell input to: and common good, definition of as unfettered individual access to cures, 30–31; location of CIRM building and, 32–39, 194n23; and marketing of stem cell research, 43–47; and minority stakeholders, efforts to mollify, 7–10, 39–42; open meeting laws, CIRM exemptions from, 31–32, 37; and patient advocates as de facto public, 69–70, 71–72, 75, 203n52. *See also* isolation of CIRM; "the people," construction of

public institutions: abandonment of by social elites, 18; and commercialization of science, 3

public pact with science and medicine, fragility of, 10

race: biological importance of, CIRM ethics committee debate on, 101–8; as proxy for genetic variability, 142–44. *See also* racial profiling, medical

racialized medicine paradigm, CIRM acceptance of, 105

Index 245

Rowley, Janet, 105, 108
Roxanne (sickle cell patient), 125
Roy, Arundhati, 171
royalty payments from biotech companies: impact on participation in CIRM, 161; legislators' efforts to obtain, 160–64; for mitigation of social inequalities, 30; as obstacle to scientific progress, 70, 162–63; protection of biotech industry from, 39; Winickoff on limitations of, 182

Sacramento, as candidate for hosting of CIRM, 33, 35
San Diego, as candidate for hosting of CIRM, 33
San Francisco, CIRM building in, 33–34
San Francisco Chronicle, 48
Sankofa bird, 25; as model for inquiry into scientific progress, 23–26, 153, 170
Saward, Michael, 151
Schwarzenegger, Arnold, 1–2, 36, 164, 199, 221–22n14
science: Big Science's ignoring of ethical and social research, 173; people's engagement with, civic epistemologies and, 153, 219–20n42; ready-made, vs. in-the-making, 22–23, 26. See also commercialization of science
science, governance of: by elites, 167–68; inclusion of wide-ranging and dissident voices, 176–77; limited social engagement in, 168. See also popular governance of science and medicine
Science in Action (Latour), 22–23
science-phobia, perception of in African Americans, 115

scientific progress, unintended consequences of, need to deal with, 26
scientists: careerism of, 149; and objectivity, xi; views on popular governance of science and medicine, 4, 178
scientists, autonomy of: commercialization of science and, 3–4; recognition of problems with, 3; stigmatization of opponents of, 51. See also popular governance of science and medicine; right to research amendment
SCNT. See somatic cell nuclear transfer
SCR. See stem cell research
self-determination. See free choice
Senate Bill 18, 161, 221n14
Senate Bill 401, 161, 164
Senate Bill 1260, 99, 221–22n14
Senate Bill 1565: as insulation from potential increases in health disparities, 39; opposition of stem cell advocates to, 37–38, 43; provisions of, 164
Senate Constitutional Amendment 13, 161, 222n14
Sennett, Richard, 1
Shavonne (egg donor), 80–81
Sheehy, Jeff, 45–46, 132
Shestack, Jonathan, 94–95
sickle cell disease: cure for, as marketing point for Proposition 71, 129–31; lack of CIRM research funding for, 131–33; lack of representation for on ICOC, 130–31; and patient noncompliance, 147–48; perceived efficacy of home remedies over medical treatment, 121–24; research funding increases for, 128; variations in severity of, 120–21

distrust of clinical research as rational response to, 138–41, 144–47, 152–53, 155; separation of groups as norm in, 169; and technology, ideology inherent in, 177; and technology, perception of as autonomous force, 178–79; traditional, breakdown of, and new symbolic realities, 219n41

socioeconomic privilege, biological citizenship model and, 51

somatic cell nuclear transfer (SCNT): and eggs, need for, 82; risks to egg donors and, 84

Somers, Nancy, 126–27

Somers, Teri, 171–72

Southeast Asians, distrust of health care system, 145

Spinal Cord Injury Association, 77

Stanford Center for Biomedical Ethics, 76

state initiatives on stem cell research: increase in, 188n18; and popular governance of science and medicine, 4–5, 12

State of Stem Cell Advocacy conference (2008), 43–47, 72

state oversight: independence of CIRM funding from, 34, 192n55; location of CIRM building and, 34–35; right to research amendment and, 18–19, 21

state power, coopting of by stem cell research movement, 18

Stem Cell Action Network, 72, 203n54

stem cell advocates: able-ist paradigm of, 64, 71; belief in universal benefits of stem cell cures, 60; biocapitalism of, 70; and disability advocates, limited influence of, 74–75; emphasis on

biological basis of conditions, 59–65, 74; emphasis on consumer rights, 72, 74; and erosion of disability activists' gains, 63–64; individual self-determination as implicit value of, 15, 17, 190n42; lobbying by, 59–60; perception of religious conservatives as main opposition to, 74–75; as supporters of quality-of-life resources for disabled persons, 61; views on disability activists, 62; views on opponents, 43–44, 51, 53, 60; views on regulations to ensure affordable access, 31

stem cell banks, public, and potential abuse of minority populations, 127–28

Stem Cell Battles (blog), 60

stem cell research (SCR): defining of past and future states in, 23; as distraction from equitable distribution of basic health care, 12–13, 15, 17–18, 50, 76, 124, 157–58, 160; egg supply, difficulty of obtaining, 82–83; as elite priority, 158; embryonic stem cells, isolation of, 11; as field of political contention, 11; governance of by elites, 167–68; government-funded, parallels with urban renewal programs, 38–39; linking of to California pioneer ethos, 12–13; marketing of to public, 43–47, 53; novel alliances created by, 7; patient advocates as de facto public for, 69–70, 71–72, 75, 203n52; social justice concerns as obstacle to, 31, 35, 49, 50, 53–54; whites as primary benefiters from, 49

stem cell research, state initiatives on: increase in, 188n18; and popu-